ERDC/CHL TR-11-10
December 2011

Verification and Validation of the Coastal Modeling System

Report 2: CMS-Wave

I0393632

Lihwa Lin, Zeki Demirbilek, Rob Thomas, and James Rosati, III

Coastal and Hydraulics Laboratory,
U.S. Army Engineer Research and Development Center
3909 Halls Ferry Road,
Vicksburg, MS 39180-6199

Report 2 of a series

Abstract: There are four reports documenting the Verification and Validation (V&V) of the Coastal Modeling System (CMS): an executive summary, waves, circulation, and sediment transport/morphodynamics, respectively. This is the second technical report (Report 2) that describes the wave modeling component of the V&V study. The goal of the report was to critically assess both general and special predictive skills of CMS-Wave, a spectral wave model in the CMS developed to address a variety of needs for coastal inlet applications. For model verification, a number of simple and idealized cases were selected to approve the basic physics and computational algorithms implemented in CMS-Wave. For model validation, a collection of more complicated cases with either laboratory or field data representing real world problems were assembled to confirm the overall performance or special capabilities of CMS-Wave. Provided in this report are descriptions of the V&V cases, model set up and boundary conditions specified in each case, and assessment of model performance. Major findings for each case are provided as guidance to users for future applications of CMS-Wave.

Contents

Figures and Tables

Figures

Tables

Preface

This study was performed by the Coastal Inlets Research Program (CIRP), funded by the U.S. Army Corps of Engineers, Headquarters (HQUSACE). The CIRP is administered for Headquarters at the U.S. Army Engineer Research and Development Center (ERDC), Coastal and Hydraulics Laboratory (CHL) under the Navigation Systems Program of the U.S. Army Corps of Engineers. James E. Walker is HQUSACE Navigation Business Line Manager overseeing CIRP. Jeff Lillycrop, CHL, is the Technical Director of the Navigation Systems Program. Dr. Julie Rosati, CHL, is the CIRP Program Manager.

CIRP conducts applied research to improve USACE capabilities to manage federally maintained inlets and navigation channels, which are present on all coasts of the United States, including the Atlantic Ocean, Gulf of Mexico, Pacific Ocean, Great Lakes, and U.S. territories. The objectives of CIRP are to advance knowledge and provide quantitative predictive tools to (a) support management of federal coastal inlet navigation projects, principally the design, maintenance, and operation of channels and jetties, and reduce the cost of dredging, and (b) preserve the adjacent beaches and estuary in a systems approach that treats the inlet, beaches, and estuary as sediment-sharing components. To achieve these objectives, CIRP is organized in work units conducting research and development in hydrodynamics, sediment transport and morphology change modeling, navigation channels and adjacent beaches, navigation channels and estuaries, inlet structures and scour, laboratory and field investigations, and technology transfer.

For the mission-specific requirements, CIRP has developed a spectral wave model CMS-Wave specifically for inlets, navigation and nearshore project applications. The model is part of the Coastal Modeling System (CMS) intended to simulate nearshore waves, flow, sediment transport, and morphology change affecting planning, design, maintenance, and reliability of federal navigation projects. The verification and validation of CMS-Wave are conducted in the assessment to determine model capability and versatility for Corps projects. The validation of CMS-Wave includes real data collected from the field and laboratory to determine reliability of wave predictions.

This report was prepared by Dr. Lihwa Lin, Coastal Engineering Branch (CEB), Dr. Zeki Demirbilek, Harbors Entrances and Structures Branch, Robert Thomas, CEB, and James Rosati III, CEB of ERDC-CHL, Vicksburg, MS. The work described in the report was performed under the general administrative supervision of Dr. Jeffrey Waters, Chief of Coastal Engineering Branch, Dr. Jackie Pettway, Chief of Harbors Entrances and Structures Branch, and Dr. Rose M. Kress, Chief of Navigation Division. Dr. Julie Dean Rosati, Dr. Jane M. Smith and Alison S. Grzegorzewksi performed the internal review of this report. Professors Okey Nwogu, University of Michigan and Vijay Panchang, Texas A&M University-Galveston, and Dr. Wei Chen, of Worley Parsons Inc, provided external reviews. Donnie F. Chandler, ERDC Editor, ITL, performed the final review and format-edited the report. Jose Sanchez and Dr. William D. Martin were Deputy Director and Director of CHL during the study and preparation of the report, respectively.

COL Kevin Wilson was ERDC Commander. Dr. Jeffery Holland was ERDC Director.

Unit Conversion Factors

Multiply	By	To Obtain
degrees (angle)	0.01745329	radians
feet	0.3048	meters
miles (nautical)	1,852	meters
miles (U.S. statute)	1,609.347	meters
square feet	0.09290304	square meters
square miles	2.589998 E+06	square meters
knots	0.5144444	meters per second
miles per hour	0.44704	meters per second

1 Introduction

1.1 Purpose

This report describes the Verification and Validation (V&V) study for the production version of CMS-Wave, a directional spectral wave transformation model. The study compared model calculations with analytical solutions and empirical relationships as well as laboratory and field data for a number of coastal applications. Presently, CMS-Wave exists in two versions: a) a stand-alone model, which is the official wave model that can interact with the explicit and implicit versions of the CMS-Flow models, and b) a research version of the model embedded within the implicit CMS-Flow. The objective of the present report is to evaluate the standalone CMS-Wave for different types of wave prediction problems in coastal applications, provide guidance for field applications, and identify potential improvements to CMS-Wave features and computational capabilities. This report augments the previous CMS-Wave technical report (Lin et al. 2008) by providing V&V test cases for existing and new features in CMS-Wave. This Chapter describes the V&V plan outlined by the Coastal Inlets Research Program (CIRP) to evaluate various capabilities of the CMS-Wave, including model set up, input forces, computational parameters, model-data comparison, and user guidance for application in navigation projects.

1.2 CMS-Wave

The USACE maintains a large number of coastal structures in support of federal navigation projects nationwide. Many of these structures, which are aging, are exposed to waves, tides and storm surge. The controlling currents and waves promote scouring of navigation channels and stabilize the location of inlet channels and entrances. Navigation structures protect vessels in transit through ship channels and inlets. Consequently, coastal navigation structures are subject to degradation from the continual impact of currents and waves impinging upon them.

Questions arise about the necessity and consequences of engineering actions taken to design, rehabilitate and modify navigation structures. District projects that support federal navigation projects and the long-term maintenance and rehabilitation of navigation structures require a continual (life-cycle) forecast of waves and currents. Potential effects of global warming and relative sea level change over the life-cycle of structures must also be considered in the analyses and decisions. CMS-Wave was developed to provide wave estimates required in these applications to planners, engineers and decision-makers.

CMS-Wave may be used as a stand-alone model or as part of the integrated CMS package. The package consists of multi-dimensional numerical models that simulate the combined effects of wind, waves, currents, water level, sediment transport and morphology change in the coastal zone. CMS-Wave is designed for the wave modeling needs of navigation projects in channels, coastal inlets, wave-structure interaction and sediment management at inlets, adjacent beaches and estuaries. To facilitate the application of CMS-Wave, the model has a user-friendly interface in the Surface-water Modeling System (SMS).

CMS-Wave is a two-dimensional (2D) spectral wave model with energy dissipation and diffraction terms (Mase et al. 2005a; Lin et al. 2011; Lin et al. 2008; Demirbilek et al. 2007a and 2007b). It performs steady-state spectral transformation of directional random waves co-existing with ambient currents in the coastal zone. The model simulates half-plane and full- plane wave propagation , so that wave generation, wave reflection and bottom frictional dissipation of multi-directional waves can be considered. Frequently questions are raised about the choice of a wave model appropriate for the needs of project. The following explanation about two classes of wave models is intended to help address this question.

There are two classes of coastal wave models used in practical applications: phase-averaged spectral wave models and phase-resolving frequency or time-dependent wave models. The CMS-Wave is a spectral wave model belonging to the phase-averaged class (Lin et al.2008; Smith et al. 2001, Booij et al. 1999). This type of wave model is used widely because of its computational efficiency for transforming incident wave conditions to the project site, for developing estimates of wave parameters in the nearshore for engineering studies, and for providing wave input to the circulation (flow) models, which can be then used to perform sediment transport and morphological calculations. Numerical wave models in this category are ideal for generation, growth and transformation of wind-waves over large distances (fetches) in regional-scale applications. CMS-Wave is an essential component of the CMS package because it has several unique features developed specifically to accommodate navigation structures at coastal inlets.

In contrast, the phase-resolving models (Demirbilek and Panchang, 1998; Panchang and Demirbilek 1999; Demirbilek et al., 2008; Nwogu and Demirbilek, 2001) have been developed to investigate the detailed patterns of complex nonlinear wave processes occurring on smaller length scales in the vicinity of coastal structures, inlets and navigation channels, the complicated evolution of waves within enclosed areas of ports / harbors / marinas, and in open coast areas with strongly varying bathymetry and irregularly shaped coastlines. These models employ much finer spatial and time-scales in intermediate and shallow waters, which are required for accurate representation of sub-wavelength wave processes (e.g., wave reflection, diffraction, breaking, dissipation, runup and overtopping, wave-wave interaction). While spectral wave models can be applied to larger domains, the phase-resolving wave models are generally used for smaller domains because they are computationally demanding. An important distinction between the two classes of wave models is that phase-resolving wave models solve for the mass and momentum conservation equations to determine the temporal and spatial changes which occur in wave height, period and direction as well as wave phase, while spectral models solve for the conservation of wave energy/action, and changes in phases of the propagating waves are not considered by these latter models (phase is considered random).

Both classes of wave models are used widely in coastal engineering, but each class has certain limitations and advantages. For example, harbor resonance (seiching oscillations) problems and wave reflection problems caused by wave-boundary interactions cannot be modeled with the wave action (or energy) conservation equation used in spectral wave models. Wind input is not considered in the governing equations of phase-resolving wave models because there is not enough time or distance (fetch) for wind input to influence wave growth and transformation over the short spatial and temporal scales over which this class of model is typically applied (e.g., winds would need longer times and greater distances to modify the input waves).

Wave processes such as wave diffraction, reflection, breaking, runup and overtopping, nonlinear wave-wave and wave-current interactions can be represented accurately by numerical models which use the conservation of mass and momentum principle. These processes are not easy to quantify accurately with spectral wave models which use linear transformation of wave energy at scales comparable to or greater than wavelengths of wind-generated waves or without phase information. Although spectral models may not accurately model highly nonlinear transformation processes of shallow water waves, these models are used frequently in coastal works because they are computationally efficient for large spatial domains and provide sufficient accuracy for engineering studies. It is possible to treat some nonlinear wave processes in an approximate fashion in the spectral models. Approximations to some nonlinear processes are included in CMS-Wave to provide improved estimates to project applications. Additional information about general features of different types of numerical wave models is available in a review article (Panchang and Demirbilek 1999). Because these two classes of wave models complement each other, they are often used in tandem to address the different needs of coastal projects.

The test examples considered in this V&V study include a wide variety of problems pertaining to practical coastal applications. Findings from the CMS-Wave application in these tests indicate that the model is applicable for propagation of random waves over complex nearshore bathymetry where wave refraction, diffraction, reflection, and breaking occur simultaneously. The details about CMS-Wave model theory and numerical implementation are available (Lin et al. 2011; Lin et al. 2010; Lin et al. 2008;Demirbilek et al. 2007a), so the emphasis in this report is strictly on the model's V&V. Model setup, inputs and outputs, comparison to data and application guidelines are described for each test case. More specifically, this CMS-Wave verification and validation study is focused on evaluation of the following capabilities of the model:

- Wind Wave Generation and Growth
- Wave Transformation and Shoaling
- Wave Diffraction
- Nonlinear Wave-Wave Interaction
- Wave and Tidal Current Interaction
- Wave Runup on a Plane Slope
- Wave–Structure Interaction
- Wave-Island Interaction
- Wave Propagation over a Reef
- Storm Wave Prediction

Some of these model capabilities may require users to specify some parameters in the model input files, while others may not. Therefore, a summary of CMS-Wave control parameters and their default values are provided in Appendix A.

1.3 Verification and Validation (V&V)

1.3.1 Need

The fundamentals of a generic numerical model V&V process are summarized in this section. A short summary is given here, and additional information about specifics of the process is available in Report 1 of this series, and related publications listed in the References section of this report (e.g., AIAA 1998; ASCE 2008; Bobbit 1988; Lynch and Gray 1978; Lynch and Davies 1995; Oberkampf and Trucano 2002; Oreskes et al. 1994; Roache 1989, 1997, 1998, 1999; Trucano et al. 2003; Wang 1994). The basic V&V approach, process and terminology used for the CMS V&V follow the outline of the American Society of Civil Engineers' (ASCE, 2008) V&V protocol closely.

Numerical models for coastal engineering applications must be capable of producing reliable estimates (e.g., reasonable agreement between models and data). In addition, model performances must be robust, i.e., the physics investigated should be reproduced consistently regardless of different sites or conditions. Various capabilities of a numerical model to predict realistic physical processes and phenomena should be confirmed before it is applied to real world projects. This is necessary because a numerical model is a complex system of equations involving unknown boundary forcings and conditions and, often, a set of input parameters that are unknown or have inherent errors. The process that evaluates a model's general skills, including its mathematical and computational capabilities by thoroughly checking, calibrating and validating the model with analytical solutions or data, is called the model V&V. The V&V is the process by which models become accepted for engineering applications. There is no assurance that a model has all the right capabilities even if it has been proven to be mathematically correct, until each of its computational capabilities has been examined, calibrated and validated with analytical solutions or data. This is the ultimate purpose of a model V&V study. The need for model Validation arises when it is necessary to quantify the confidence in the predictive capability of the model's computational code through comparison with a set of data. However, several technical challenges must be addressed for a successful Validation of a model's predictive capability. One of these challenges is "what is an acceptable model-data discrepancy?" There is no formal definition or metric that can be used for a "model discrepancy" for different types of coastal engineering models, and this is to be decided by the person conducting the evaluation. Another challenge involves training and persuading the coastal engineering community about model suitability and the calibration and validation paradigms given the uncertainties involved in both the model and data.

1.3.2 Significance of V&V in Engineering and R&D

V&V has gained more attention in recent decades because of increasing use of numerical models in engineering practice. Numerical models are simplified representations of reality which are useful for guiding further research and advances in science, but are not proof of reality. Since numerical models go beyond the range of available analytical solutions, they cannot be verified in applications. One would have a greater confidence while applying a model to a field study if it were already fully verified with analytical solutions, empirical formulas or physical model data.

Engineers use Verification to elucidate discrepancies in other models by performing sensitivity analysis for exploring "what if" questions. Verification can also help researchers to identify which aspects of a numerical model need further study or where additional empirical data are needed. Sometimes the term "bench-marking" is used for verification that best describes the comparison of numerical models to analytical solutions, since it denotes a reference to an accepted standard whose absolute value is unknown.

For researchers and practioners, Validation is defined as a demonstration of proof that a model within its domain of applicability possesses a satisfactory range of accuracy consistent with the intended application of the model. The Validation process for coastal wave models is multi-faceted as it involves several levels of checks and analyses. Some consider the Validation to be a subjective evaluation because it is often defined in terms of an "acceptable degree" or a "reasonable agreement" between the model and data, the terms which are commonly used in model performance. Such subjective measures can make model validation a debatable issue if reproducibility is not guaranteed. In this V&V study, when applicable, statistical error metrics have been used to quantify the true differences between the model and data to avoid the issue of a subjective assessment.

In practice, the adequacy of a model matching data is the only true measure of the model's validity and acceptance to be used in projects. Complex coastal wave models are likely to be valid for some predictions they produce and invalid in others. However, not all models need to be validated, and the chosen level of validation depends on the model's purpose. Complex wave models can never be fully valid because of their extensive features and uncertainties in the model's settings and input conditions. A partial validation of model's features helps only to make it to be a "qualified" tool for engineering applications, but this does not assure complete validity, suitability or adequacy of the model for all project needs. Validation always involves a choice of spatial and temporal scales and only the dominant physics. Consequently, a model can be validated only for a certain range of conditions. Since verification and validation are technical terms, it is better to avoid using any generic metrics when quantifying the verification or validity. Degrees of acceptability and indices must be used to describe the true performance of a model objectively.

1.3.3 Approach

Uncertainties are inherent both in data and in models. Consequently, a sensitivity analysis is essential to understand the effect of various uncertainties, and becomes important in a numerical model's calibration and validation. The simplest model calibration is done by adjusting a set of parameters associated with computational features of the code in an attempt to maximize the model and data agreement.

Data used in the model V&V are from two main sources: physical models and field experiments. Analytical solutions provide an opportunity to test the basic physics implemented in a model. Because wave conditions in physical models are controlled and physical processes are observed with minimal uncertainty, repeated agreement between a numerical model and laboratory measurements is a good indicator of the consistency of true physics, although some of the physics may not exist in the laboratory studies. Nonetheless, model and laboratory data agreement is not sufficient evidence that the model would perform well in a prototype environment because prototype contains additional physics. Laboratory scaling effect is also a concern. Given that a coastal wave model has many capabilities, validation using a large set of physical models and field data would be necessary to evaluate the model's different features which may depend on the temporal and spatial scales of problems. Because both laboratory and field data have certain advantages and disadvantages, a combination of them has been used in the present V&V study. These benchmark evaluations provide a solid first step in the assessment of CMS-Wave for field-ready applications.

There are two types of model evaluations that can be done: Verification and Validation. Verification and Validation are intended for different purposes, although they are often used interchangeably in the engineering realm. Both of these methods include a quantitative model-to-data intercomparison or model-to-model intercomparison. Both evaluations involve assessment of the methods and data required for implementing such intercomparisons meaningfully. In this V&V study, the aim of the Verification process is to answer some fundamental questions about the numerical model which include: are the right equations used in the model; are the governing equations implemented correctly and solved properly and accurately in the model; does the model solution converge? In contrast, the Validation process is intended to answer one important question: do the governing equations represent field data accurately? These differences between Verification and Validation processes are important in understanding the work described in this V&V study.

Performance of numerical models is judged by the demonstration of agreement between data and model results, also called calculated results. Because models use a simplified representation of some complex phenomena, they can only be evaluated in relative terms because models require input parameters that are incompletely known. Numerical models are often used to speed up the process of finding a "good enough" solution, where an exhaustive search for a true solution may be too costly or impractical.

Scientists and engineers often use analytical solutions to a boundary value or initial value problem to verify models. This verification includes a comparison of a numerical solution with an analytical one to demonstrate that the two match over a particular range of conditions under consideration. These comparisons are essential steps in numerical code development. The failure of a numerical code to reproduce an analytical solution should be cause for concern. However, even a full agreement between numerical and analytical solutions does not guarantee the correspondence of either one to reality. A numerical solution verified in the realm of an analytical solution cannot be considered truly verified beyond the range and realm of the analytical solution. Consequently, numerical model validation using the physical model data is an essential next step to confirm if a particular model is capable of reproducing true physical processes and mechanisms. The model can be applied to study real world problems only after it is validated. A revalidation would be necessary as a new mechanism in when an application is concerned.

1.3.4 Definitions

For the purpose of this study, a formal definition of a numerical model, as well as what we mean by Verification and Validation processes are provided. Introducing this formalism should help readers to understand clearly the usage of some technical terms which can be controversial and subject to other definitions. These definitions apply only to computational solutions of partial differential equations generated by finite difference, finite element, spectral, or other numerical methods. Based on the above background and discussion of semantics involved, the definition of some key terms used in this V&V study is given below:

Code is the software that implements the solution algorithms. In other words, a computer code here refers to the software implementation component of a numerical model. A coastal numerical modeling system involves:

1. adapting partial differential equations with initial and boundary conditions;
2. developing mathematical algorithms for the numerical solution of equations;
3. implementing these algorithms in a computer software package;
4. executing the code on personal computers, and
5. analyzing the results produced by the modeling system.

Verification is the process of determining that a model implementation represents the developer's conceptual description of the model and the solution to the model accurately. In other words, verification is the process of confirming that a computer code implements the algorithms that were intended correctly. This is done by comparing its solutions to analytical solutions and empirical formulas confirmed by limited data.

Validation is the process of determining the degree to which a model is an accurate representation of the real world from the perspective of the intended uses of the model. In other words, it is the process of confirming that the calculations represent measured physical phenomena adequately. This is done by comparing to physical model and field data.

With the above definitions, the output of the executed code is referred to as the result or calculation, obtained for a given input used to execute the code. Engineers also use the term model calibration, which simply means adjusting a set of code input parameters usually associated with some aspect of physics that at present cannot be fully described. The purpose is to maximize the agreement between the code's calculations and data, which is generally expressed as a quantitative specification of the agreement. Calibration is not the same as validation, which quantifies our confidence in the predictive capability of a code for a given application through comparison of calculations with a set of data collected from a laboratory or the field. Given these definitions, four observations can be made:

a. Validation and calibration depend on results of verification,
b. Calibration should use the results of prior validations,
c. Calibration cannot be viewed as an adequate substitute for validation in engineering applications, but can be a good first step for future applications under similar conditions,
d. Validation should occur after calibration and not use the same data.

This report will not address the quality of data used in the verification and validation. The type of data and instrument accuracy can be found in the references provided. In general, data collected in a laboratory experiment are more reliable than data obtained from a field site because there is less uncertainty about test conditions used to gather data in a controlled physical modeling environment. Consequently, the statistical comparisons of a model with data from laboratory experiments provide a better measure of assessing the model's process-specific skills. Comparisons to field data shed more light into the model's ability to replicate combined wave processes existing in prototype environment. Furthermore, because CMS-Wave is a steady-state model, it is ideally suited to laboratory studies given that in the real world, the steady-state or equilibrium condition seldom exists.

1.4 Study plan

The test cases used in this V&V study were grouped into three categories:

1. Analytical solutions, empirical formulas, and idealized problems (Verification),
2. Laboratory studies (Validation), and
3. Field investigations (Calibration and Validation).

Each category consisted of examples representing known analytical solutions or empirical formulas for idealized problems, and laboratory and field studies with data available from the physical model or field experiments. The numerical modeling test cases are listed in Table 1. Others will be investigated and the results presented in future companion reports. Later in this report, the purpose of each test case will be described, with a discussion of which wave processes are checked with each selected test case, and a quantitative model performance will be provided. Only a list of the V&V test problems which have been completed to date follows.

Table 1. V&V test cases for CMS-Wave.

Processes Involved	Category 1: Analytical/Empirical Solutions	Category 2:Laboratory Applications (with data)	Category 3: Field Applications (with data)
Wind-wave generation Propagation in half plane Propagation in full-plane	CEM/SPM curves (for wave generation and growth over short, long and fetch-limited distances)		Matagorda Bay, TX Mouth of Columbia River, WA/OR Ship Island, MS (MsCIP)
Wave-wave interaction Infragravity waves	Idealized JONSWAP case		Mouth of Columbia River, WA/OR
Wave breaking formulas Wave-current interaction		Smith et al. (1998) idealized inlet experiments Visser (1991) experiments	Field Research Facility, Duck, NC Mouth of Columbia River, WA/OR Grays Harbor, WA Matagorda Bay, TX Southeast Oahu, HI Indian River County, FL
Wave diffraction Wave reflection	CEM/SPM diffraction curves (gap problem)		Grays Harbor, WA
Wave-structure interaction Wave runup Wave transmission Wave overtopping		Ahrens and Titus (1981), Ahrens and Heimbaugh (1988), Mase and Iwagaki (1984), and Mase (1989) experiments Demirbilek et al. (2010)	Mouth of Columbia River, WA/OR Grays Harbor, WA

1.4.1 Category 1: Basic verification for idealized problems

Test cases in this group were used to check CMS-Wave features for the following processes:

a. Test for wave generation and growth over long- and short- fetches and fetch-limited conditions. Model calculations were compared to an analytical or empirical solution available in the Coastal Engineering Manual (CEM; USACE 2006) and the Shore Protection Manual (SPM, USACE 1984).
b. Test for wave-wave interactions for idealized JONSWAP spectrum propagation over long distance with wind input. CMS-Wave calculations were compared qualitatively to results available from the open literature (Jenkins and Phillips 2001)).
c. Test for wave diffraction at a breakwater and breakwater gap. CMS-Wave was compared to the CEM/SPM monograms.

1.4.2 Category 2: Laboratory studies with data

This group includes the following laboratory studies:

a. CHL Idealized Inlet physical model experiments (Smith et al. 1998). Test CMS-Wave for wave-current interaction and wave breaking.
b. Wave breaking experiments on a planar beach (Visser 1991). Evaluate CMS-Wave calculations for wave breaking on smooth and rough-surface slopes.
c. Wave runup over sloping structures (Lin et al. 2008). Compare CMS-Wave calculations for wave runup on various slopes.
d. Cleveland Harbor experiments (Bottin 1983). Evaluate CMS-Wave for wave-current interaction, wave diffraction, wave reflection, and wave transmission.
e. University of Delaware experiments (Chawla and Kirby 2002). Test CMS-Wave for monochromatic and random wave breaking.
f. CIRP idealized inlet physical model experiments (Seabergh et al. 2002 and 2005). Compare CMS-Wave calculations to measurements that include wave-structure interaction.
g. Large-Scale Transport Facility (LSTF) artificial reef experiments (Smith 2011). Evaluate CMS-Wave calculations for wave breaking and transmission over a reef.

h. University of Michigan wave runup over a fringing reef experiment (Demirbilek et al. 2007b). Evaluate CMS-Wave calculations for wind forcing, wave breaking and transmission on a reef bottom.

i. LSTF wave-induced longshore current experiments (Hamilton and Ebersole 2001). Test CMS-Wave for wave breaking and wave-current interaction.

j. Wave Transmission over Breakwaters (Goda 1985). Compare CMS-Wave calculations to data including wave-structure interaction.

1.4.3 Category 3: Field studies with data

This group includes test examples selected from the following field applications:

a. Matagorda Bay, Texas (Puckette 2006). Test CMS-Wave for wind wave generation in a bay.

b. Grays Harbor, Washington (Osborne and Davies 2004). Test CMS-Wave for tidal current, wind, and wave in navigation channel near jetties.

c. Mouth of Columbia River, Oregon/Washington (Moritz 2005). Evaluate CMS-Wave calculations for tidal current, wind, and waves in navigation channel and jetty rehabilitation.

d. Southeast Oahu coast, Hawaii (Cialone et al. 2008). Test CMS-Wave for wave propagation over a reef.

e. Recent FRF, North Carolina, wave measurements (Hanson et al. 2009). Evaluate CMS-Wave for cross-shore variation of storm waves at FRF.

f. Mississippi Coastal Improvement Program (Wamsley et al. 2011). Test CMS-Wave for wave prediction around a barrier island.

g. Indian River County, Florida (SES 2011). Test CMS-Wave for wave propagation over a rocky bottom coast.

h. Pillar Point Harbor, California. Test for wave prediction in and around Half Moon Bay Harbor[1] (HMB 2011).

i. Noyo Harbor, California. Test CMS-Wave for wave calculations in the vicinity of a dredge material placement site[2] (NH 2009).

j. Galveston Bay, Texas. Test CMS-Wave for wave prediction at the Houston-Galveston Ship Channel[3] (HSC 2010).

---

[1] (http://cirp.usace.army.mil/news/CIRP_News/CIRP_eNewsletter_Jun2011.pdf)

[2] (http://cirp.usace.army.mil/news/CIRP_News/CIRP-news-Dec09.html)

[3] (http://cirp.usace.army.mil/news/CIRP_News/CIRP-news-Mar10.html)

Table 1 provides a summary of all test cases included in this V&V study. To highlight the purpose of each test case under three categories, Table 1 also lists the key associated wave processes involved in the test case.

1.5 Report organization

This report is organized in five chapters. Chapter 1 presents the motivation, definitions, and an overview of the CMS-Wave V&V study. Chapter 2 describes the Verification of CMS-Wave with analytical solutions, empirical formulas for idealized cases (Category 1). Chapters 3 and 4 describe the Validation of CMS-Wave with comparison of model calculations to the laboratory (Category 2) and field (Category 3) data, respectively. Chapter 5 summarizes the overall study findings and outlines the future work.

2 Category 1 Test Cases: Basic verification for idealized problems

2.1 Overview

The comparison to empirical and analytical solutions for a few idealized problems is presented in this chapter to verify the CMS-Wave model, and to confirm that the intended numerical algorithms are implemented correctly in the model. Each case has an identifier with the first two characters indicating the Category number, followed by the Example number under the Category. For example, test case C1-Ex1 refers to Category 1 - Example 1. This notation is used henceforth.

The following three test cases have been completed:

1. Wave generation and growth over long- and short- fetches and fetch-limited conditions.
2. Wave-wave interactions for idealized JONSWAP spectrum propagation over long distance with wind input.
3. Wave diffraction at a breakwater and breakwater gap.

2.2 Test cases

2.2.1 Test C1-Ex1: Wave generation and growth in limited fetch

Description: The purpose of this test was to compare CMS-Wave prediction to wave generation and growth curves for fully-developed seas given in the Shore Protection Manual (*SPM* 1984) under the fetch-limited condition. The wave generation and growth curves (see SPM, Equations 3-33 through 3-38 and Figures 3-22 through 3-24) are based on the Sverdrup-Munk-Bretschneider (SMB) method, developed originally in the 1950s for deepwater wave growth and forecasting. The present SMB diagram shown in the SPM has improved for shallow-water constant depth applications. Additional depth and fetch applications are considered under Category 3 test cases (field validations).

Model setup and parameters: CMS-Wave was set up for a flat bottom seabed of constant depth 20 m (Lin et al. 2008). The model domain was 2 km x 20 km consisting of 10×100 cells with constant cell size 200 m $\times$

200 m. The test cases included a constant wind of 10, 20, and 35 m/sec with wind direction along the long axis (20 km) of the grid. The wave energy input at the upwind boundary of modeling area was set to zero. The wave generation was calculated on a spectral grid of 30 frequency bins (0.12 to 0.35 Hz with 0.008-Hz increment) and 35 direction bins (covering a half-plane with 5-deg spacing). The default parameters were used (see Appendix A).

Results and Discussion: Figure 1 shows comparison of the calculated wave height (top panel) and wave period (lower panel) with results from the SPM. Calculated wave height values agree well in these simulations for fetch greater than 5 km. For a fetch length less than 5 km at high wind speeds, CMS-Wave predicts lower wave heights than the SPM method, which is not necessarily an error since SPM method is only a coarse approximation. Over the short fetch lengths considered in this test, calculated wave heights by CMS-Wave increase linearly with the length of fetch, which is in agreement with the SPM method.

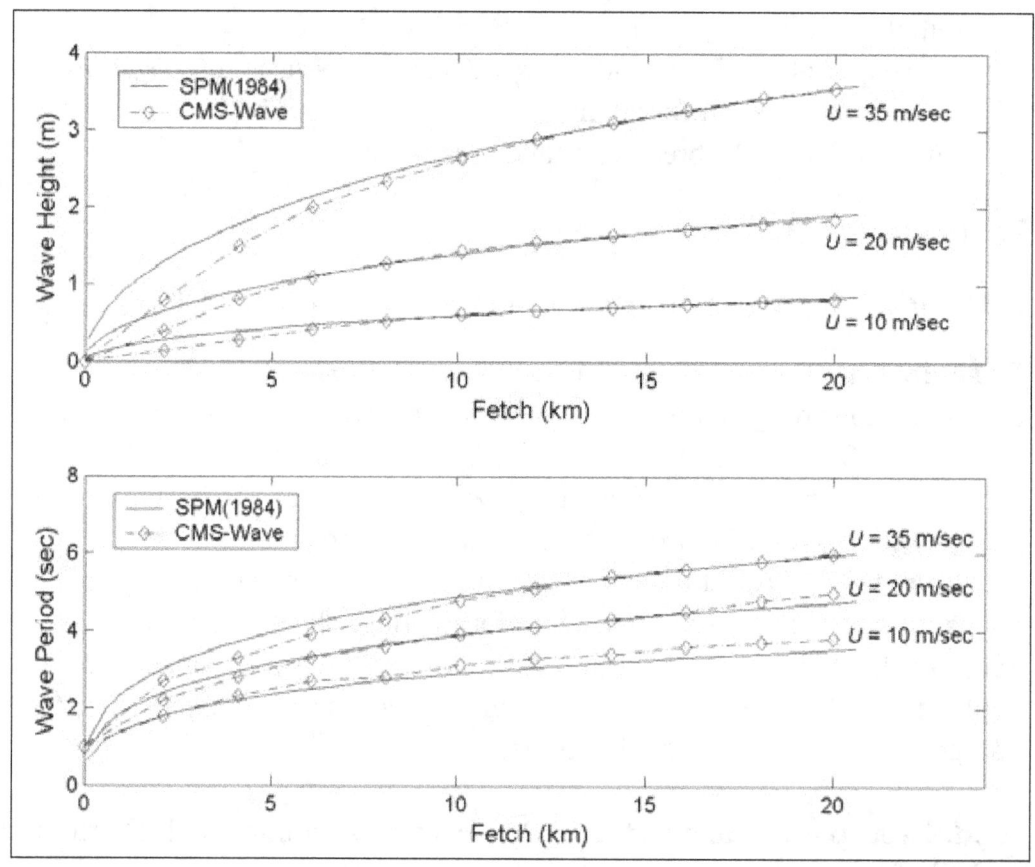

Figure 1. Comparison of calculated wave generation and SMB curves, wave height (top panel) and period (lower panel).

In summary, wave height and wave period predictions from CMS-Wave were compared to the SMB curves given in the SPM. Simulations were performed for low, moderate and strong wind speeds and for values of fetch length from 0 to 20 km. The CMS-Wave calculation and SMB curves show better agreement for a fetch greater than 5 km. CMS-Wave can calculate the wave generation and growth in the coastal and estuary area for fetches greater than 5 km. Future tests will consider shorter and longer fetches as well as different wind speeds and water depths.

2.2.2 Test C1-Ex2: Nonlinear wave-wave interactions

Description: The purpose of this test was to compare nonlinear wave-wave interactions as calculated by CMS-Wave to the analytical solution available in the open literature. Jenkins and Phillips (2001) proposed a simplified formulation to represent nonlinear wave-wave interactions as a second-order diffusion operator that conserves wave energy and wave action. Because the formulation is independent of the dispersion relation, it is applicable in both deep and shallow water. The Jenkins and Phillips formulation has been extended to both deep and shallow waters and implemented in the wave-action balance equation of the CMS-Wave to calculate nonlinear wave-wave interactions efficiently (Lin et al. 2010).

Model setup and parameters: CMS-Wave was set up with a flat seabed of constant depth 1,000 m. The model domain was 2 km x 200 km consisting of 10 × 1,000 cells with constant cell size 200 m × 200 m. The incident spectral waves included three significant heights of 0.45, 0.5, and 0.6 m with the JONSWAP (Hasselmann et al. 1973) peak enhancement parameters of 1, 2, and 5, respectively. Wind forcing was not included in these simulations. The peak period of incident deepwater wave was 10 seconds and the mean wave direction was along the long axis (200 km) of the grid. The directional distribution was a cosine bell function with a power of 20. The wave spectrum was specified in CMS-Wave on 30 frequency bins (0.04 to 0.33 Hz with 0.01-Hz increment) and 35 direction bins (covering a half-plane with 5-deg spacing). The default parameters given in Appendix A were used for this test case.

Results and Discussion: Figure 2 shows the comparison of directionally integrated wave energy transfer rate S_{nl} from CMS-Wave with exact computations in the examples of Hasselmann et al. (1985) for the spectral peak enhancement parameter $\gamma = 2$ and $\gamma = 5$. The calculated

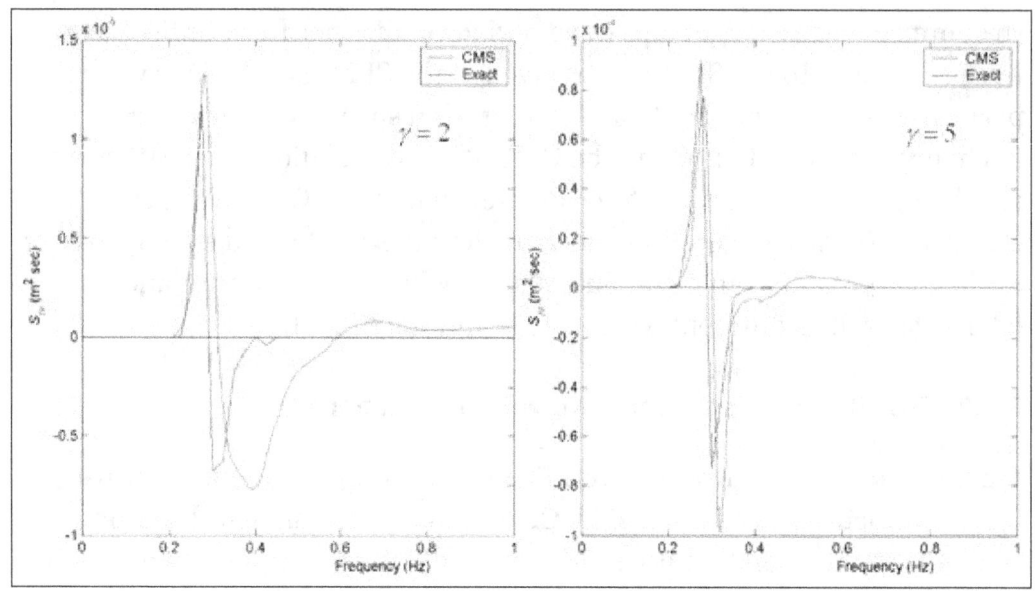

Figure 2. Comparison of directionally integrated wave energy transfer rate S_{nl} for JONSWAP spectrum with the spectral peak enhancement factor $\gamma = 2$ and 5.

results are consistent with the observed and theoretical results in that the nonlinear wave-wave interactions cause wave energy to transfer from high to low frequencies, a well-accepted process referred to as frequency downshifting. However, the simplified formulation did not fully reproduce energies at high frequencies, i.e., in the more dynamic and unstable range involving wave breaking and energy exchange with atmospheric forcing.

CMS-Wave can represent the frequency increases and decreases (up- and down-shifting processes) and corresponding wave energy re-distribution associated with nonlinear wave-wave interactions efficiently. More accurate calculations can be obtained with highly nonlinear models; however, these are computationally extremely demanding and cannot be used on desktop machines. The approximation implemented in the CMS-Wave is intended to address this important need in projects.

Lin et al. (2010) conducted additional tests to demonstrate that the nonlinear wave-wave interaction is more significant in large coastal domains from deep to shallow water with strong wind conditions. Future tests will evaluate this capability for large domains, different water depths, and greater wind wave conditions. In particular, more tests are needed to determine the applicability of the extended formulation to shallow depths and short fetch applications.

2.2.3   Test C1-Ex3: Wave diffraction at breakwater gap

Description: The purpose of this test was to compare CMS-Wave calculations to wave diffraction diagrams in the CEM (see Figures II-7-7 through II-7-17) and SPM (see Figures 2-40 through 2-59), based upon the Sommerfeld solution at breakwaters. These wave diffraction diagrams were compiled originally by Wiegel (1962) for a straight semi-infinite long breakwater and by Johnson (1952) for a breakwater gap for monochromatic incident waves impinging on these structures from different directions.

Model setup and parameters: The CMS-Wave computational domain was a square grid consisting of 101 × 101 cells with cell size of 20 m × 20 m (Lin et al. 2008). The breakwater was specified as dry cells in a column parallel to the seaward boundary. A uniform water depth of 1,000 m was specified to represent deepwater waves. The incident monochromatic wave height was 1 m, and the wave period was 8 sec (0.125 Hz), corresponding to a wavelength of 100 m. Diffraction was simulated in CMS-Wave on 10 frequency bins (from 0.06 Hz to 0.15 Hz with 0.01-Hz increment) and 35 direction bins (covering a half-plane with 5-deg spacing). In these numerical simulations, the incident wave energy was placed in a single frequency and direction bin. The incident wave was perpendicular to the breakwater. The default parameters were used (see Appendix A). The diffraction intensity κ was set to 4 (default value) to simulate the maximum diffraction estimate by CMS-Wave, and the bottom friction loss was neglected in these calculations.

Results and Discussion: Figure 3 shows calculated wave height normalized by the incident height (dash-dot line) and wave diffraction diagram (solid line) for a semi-infinite long breakwater. The normalized diffraction wave height is also termed as the diffraction coefficient, K'. The graph coordinates are normalized by the wavelength, and plots are in units of wavelength (L) such that the diffraction diagram can represent the deepwater wave as well as shallow to intermediate water conditions. Calculated wave diffraction coefficients agree qualitatively with the values presented on diffraction diagrams in the CEM and SPM.

Figures 4 and 5 show wave diffraction diagrams (SPM 1984) and calculated wave heights for the breakwater gap width, $B = L$, and $B = 2L$, (e.g., equal to one and two wavelengths, respectively). The incident wave direction was normal to the breakwater. Through calibration, the diffraction intensity value κ was set to 2 for the gap width $B = L$ and $\kappa = 3$ for $B = 2L$. Bottom

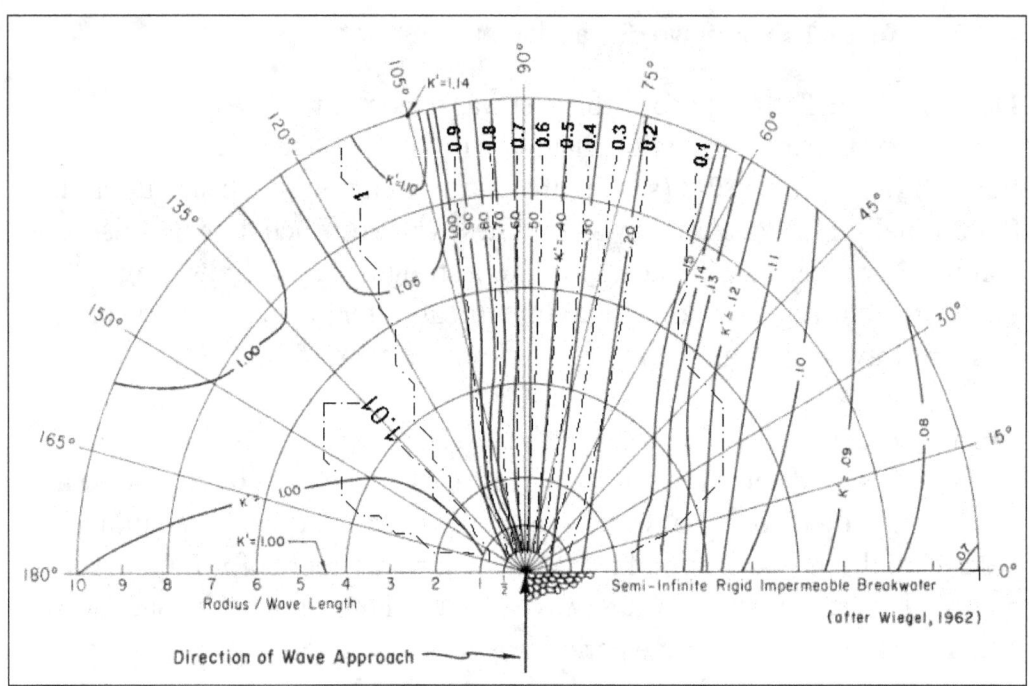

Figure 3. Wave diffraction diagram and calculated normalized diffraction wave height K'(dash-dot) for a breakwater.

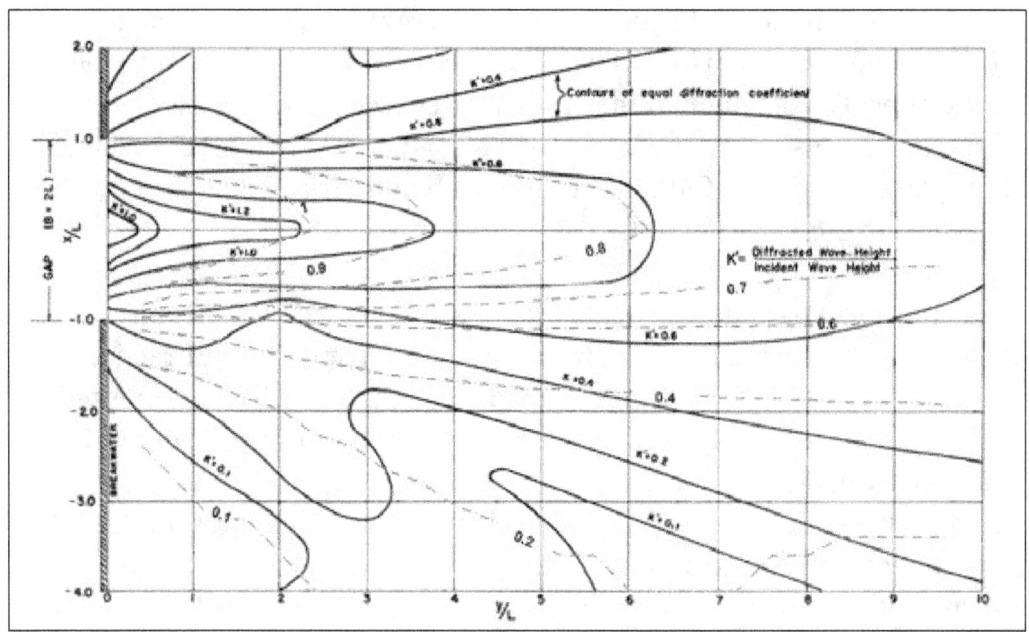

Figure 4. Wave diffraction diagram and calculated K'(dash-dot) for a gap, $B = 2L$.

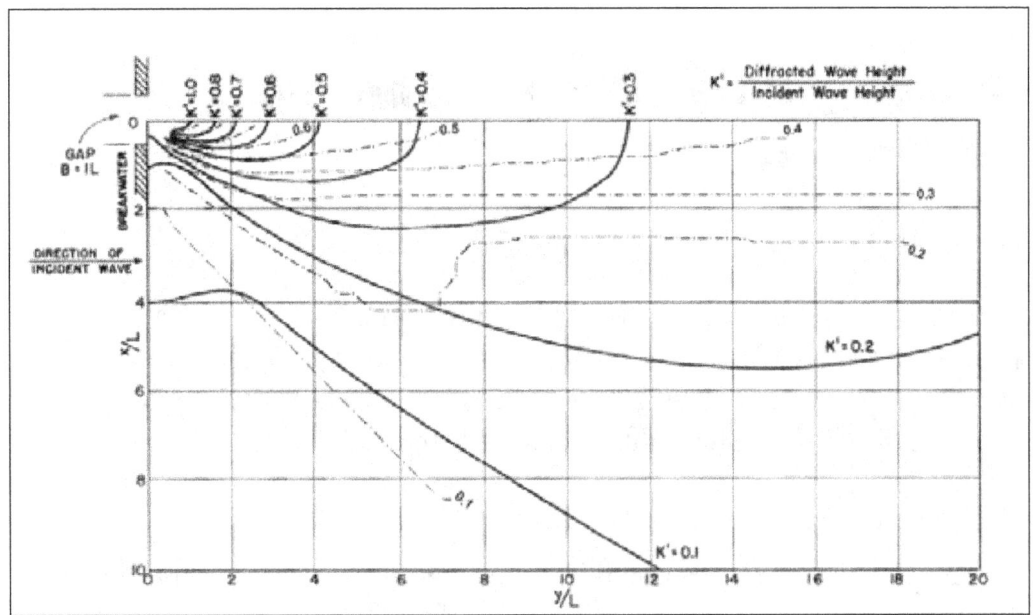

Figure 5. Wave diffraction diagram and calculated K' (dash-dot) for a gap, $B = L$.

friction was neglected. Calculated wave diffraction coefficients agree qualitatively with the CEM and SPM diffraction diagrams in these breakwater gap simulations.

In summary, wave height and direction predictions from CMS-Wave were compared to the analytical solutions given in the CEM/SPM. Simulations of diffracted waves were performed for a gap of a width B= 0, L and 2L (L=wavelength) and constant depth. Simulations were performed by placing the entire incident wave energy in one frequency and direction bin. Model predictions replicated analytical solutions closely. The agreement was better in the strong diffraction zone (0 to L distance down-wave of the gap), and as waves propagated further from the gap, the difference increased gradually between model-analytical solutions.

CMS-Wave can calculate wave diffraction for engineering practice in feasibility level studies and preliminary works. However, when greater accuracy is needed in applications, a phase-resolving wave model, such as CGWAVE, a Mild-Slope Equation model (Demirbilek and Panchang 1998), or BOUSS-2D, a Boussinesq model (Nwogu and Demirbilek 2001) , or a physical model study could be used.

3 Category 2 Test Cases: Laboratory Studies with Data

3.1 Overview

The test cases presented in this Chapter are under the Category 2 type of problems, with data from physical models. The Category 2 V&V test cases completed are listed below. The remaining cases listed below are under investigation and will be included in future reports.

Completed cases:

1. Smith et al. (1998) idealized inlet experiments. Test for wave breaking on a current at an inlet.
2. Visser (1991) experiments. Test for wave breaking on a planar beach.
3. Ahrens and Titus (1981), Ahrens and Heimbaugh (1988), Mase and Iwagaki (1984), and Mase (1989) experiments. Test for wave runup over sloping structures.
4. Demirbilek et al. (2010) wave propagation into Cleveland Harbor, Ohio. Test for wave shoaling, refraction, diffraction, wave-current interaction, and wave transmission over breakwaters.

Cases in progress:

1. Chawla and Kirby (2002) experiments.
2. Seabergh et al. (2002 and 2005) CIRP idealized inlet physical model experiments. Smith (2011) LSTF experiments for waves over artificial reef.
3. Demirbilek et al. (2007b) University of Michigan experiments for wave runup over a reef.
4. Hamilton and Ebersole (2001) LSTF experiments for wave-induced longshore currents.

Three statistical measures are used as "goodness-of -fit" error metrics to characterize the level of agreement obtained between the model and laboratory data. These are the Root-Mean-Square Error (RMSE), Correlation Coefficient (R), and Mean Absolute Error (MAE), defined as

$$\text{Root Mean Square Error:} \quad \text{RMSE} = \sqrt{\frac{\sum_{i=1}^{N}\left(x_{c,i} - x_{m,i}\right)^2}{N}} \quad (1)$$

Correlation Coefficient:

$$R = \frac{\sum\limits_{i=1}^{N}\left(x_{c,i} - \overline{x_c}\right)\left(x_{m,i} - \overline{x_m}\right)}{\sqrt{\sum\limits_{i=1}^{N}(x_{c,i} - \overline{x_c})^2}\sqrt{\sum\limits_{i=1}^{N}(x_{m,i} - \overline{x_m})^2}} \qquad (2)$$

Mean Absolute Error:

$$\mathrm{MAE} = \frac{\sum\limits_{i=1}^{N}\left|x_{c,i} - x_{m,i}\right|}{N} \qquad (3)$$

where $x_{c,i}$ and $x_{m,i}$ are the i-th calculated and measured values, respectively, in a total of $i=1$ to N samples; $\overline{x_c}$ and $\overline{x_m}$ are the mean values of $x_{c,i}$ and $x_{m,i}$, respectively. The sample as used here refers to an individual test in an experiment or to a specific gauge within a test. Furthermore, the values of "x" represent parameters of zero-moment wave height, peak or mean period, and peak or mean direction. Consequently, these are not "samples" in the sense of standard sample measured time series of the water surface in experiments or calculated spectral wave parameters, but post-processed results of those samples.

The Pearson correlation coefficient (R) in Equation (2) and coefficient of determination ($R^2 = R*R$), both dimensionless, are most frequently used in engineering works to indicate agreement between different datasets (e.g., numerical model results and data). R varies between -1 and 1 while R^2 is accordingly bounded between 0 and 1. Because R measures the linear co-variation between two datasets, higher R or R^2 indicates that the two datasets have similar linearly spatial or temporal patterns. However, the use of R or R^2 can be sometimes misleading to measure data agreement because they fail to measure the actual difference between two datasets. Consequently, neither R nor R^2 alone may not be a good measure of data agreement. Additional statistics are required to quantify the agreement between model and data. The mean bias is a simple algebraic difference between datasets $x_{c,i}$ and $x_{m,i}$ of sample size N, which measures the average difference between the two datasets. Two other meaningful statistical error measures commonly used are the MAE and RMSE. The R, MAE and RMSE metrics measure the actual differences between two different datasets, but they are not standardized and not bounded. When these error measures are expressed as percentage errors, they are standardized and are independent of the unit of data. To remedy the shortcomings of these individual metrics, Willmott (1981 and 1982) developed the index of agreement that embodies R, R², MAE and RMSE in a single expression, it is non-dimensional and bounded between 0 and 1, but is rarely used in coastal engineering practice. Among these and many other error measures which are used in engineering and science, each one has advantages and disadvantages. In this study, the R, RMSE and MAE, as defined above in Equations (1), (2) and (3), are used for "goodness-of-fit" error measures to evaluate the agreement between the model and laboratory or field data.

3.2 Test Cases

3.2.1 Test C2-Ex1: CHL Idealized Inlet Experiments

Description: Smith et al. (1998) conducted a laboratory experiment to investigate wave-current interaction and the associated wave breaking in an idealized entrance with dual jetties. Figure 6 shows the bathymetry modeled by Smith et al. (depth in cm) consisting of a steep beach, and the arrangement of wave and current meters. The dual jetties had a spacing of 3.7 m and extended 5.5 m offshore to protect the entrance channel where the depth varied from 9 cm to 12.8 cm. The inlet throat converged to a depth of 15.2 cm.

Twelve wave/current conditions were tested, Runs 1 through 12, covering a wide range of wave and current parameters. In the experiment, Runs 1 to 4 were without a current, and Runs 5 to 8 had a moderate steady-state ebb (offshore) current of approximately 11 cm/sec at the inlet entrance. Runs 9 to 12 had a strong steady-state ebb current of approximately 22 cm/sec at the entrance. All waves were generated in the basin perpendicular to the shoreline with a unidirectional plunge-type generator. The wave spectra of the TMA (Bouws et al. 1985) spectral shape was applied with a gamma value of 3.3 (spectral peak enhancement parameter). Table 2 presents the incident wave parameters (significant height H_{mo} defined as 4 times the square-root of the total energy density, spectral peak period T_p, and spectral peak frequency f_p) and the ebb current speed (U). Wave data were collected along one transect line in front of the wave maker and three shore-normal transect lines in the entrance channel (Figure 6). Only the centerline data are used in this V&V study.

Model setup and parameters: CMS-Wave simulations were run at laboratory scale. The default parameters were used (see Appendix A). The

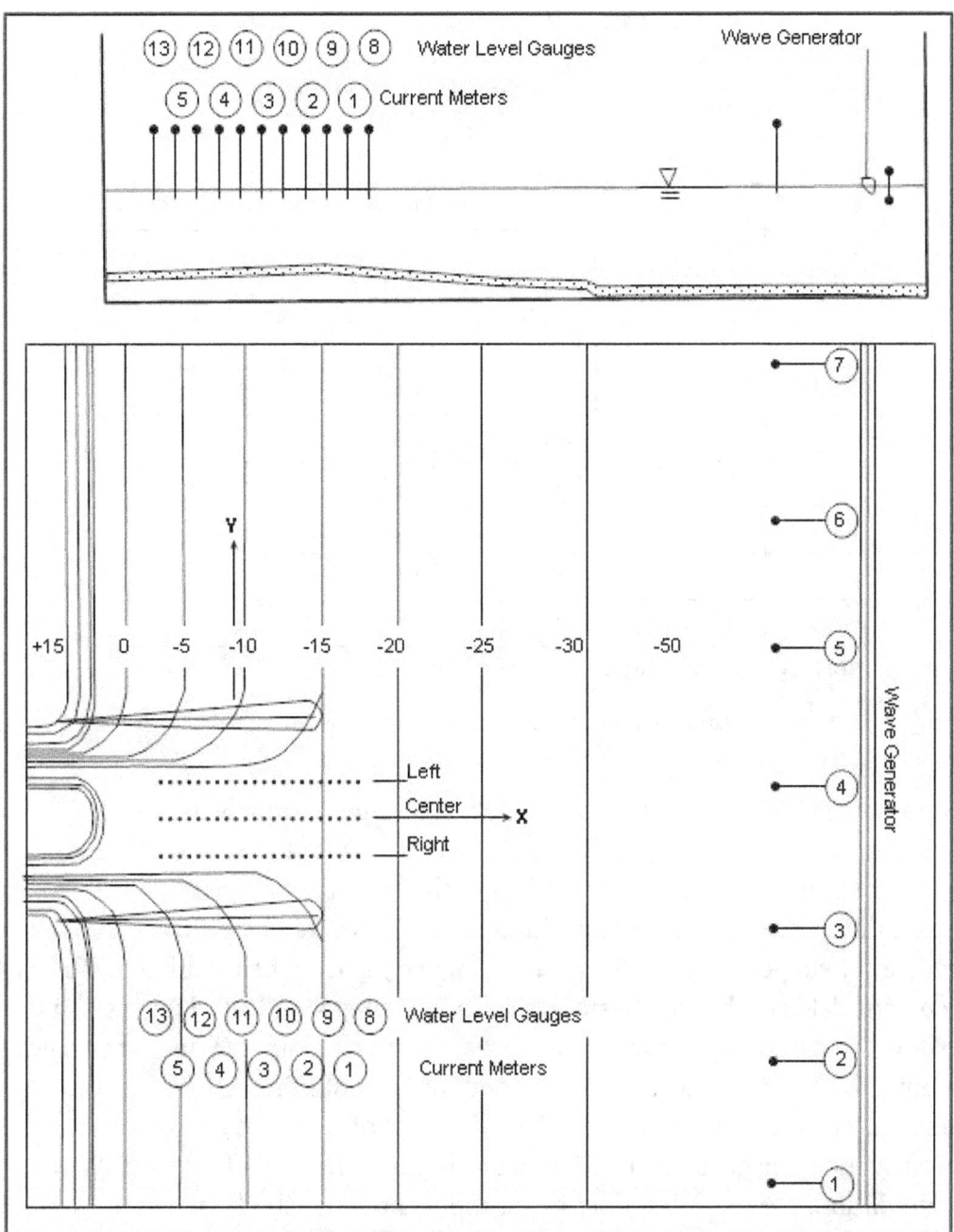

Figure 6. Idealized inlet and instrument locations (from Smith et al. 1998).

grid domain covered the same rectangular area as the experimental basin. It consisted of 188 cross-shore and 401 along-shore square cells, each 10 cm ×10 cm. The spectral wave transformation was computed using 30 frequency bins (0.5 to 3 Hz at 0.085-Hz increment) and 35 direction bins (covering a half-plane with 5-deg spacing). The 2D background input current fields for Runs 5-8 and 9-12 were prepared using CMS-Flow for different water levels specified at the inlet throat and sea boundaries. Wave

Table 2. Incident wave parameters and current conditions.

Run	H_{mo} (cm)*	T_p (sec)*	f_p (Hz)**	U (cm/sec)***
1	5.59	1.41	0.71	0
2	3.70	1.41	0.71	0
3	5.15	0.71	1.41	0
4	3.71	0.71	1.41	0
5	5.77	1.41	0.71	11.5
6	4.08	1.41	0.71	11.7
7	5.30	0.71	1.41	11.4
8	3.92	0.71	1.41	11.1
9	5.97	1.41	0.71	21.9
10	4.61	1.41	0.71	21.8
11	5.51	0.71	1.41	21.9
12	4.16	0.71	1.41	21.5

* Data averaged over gauges in front of wave generator.
** fp = 1/Tp.
*** Averaged over current meters in the entrance channel.

and flow model grids were identical. By trial-and-error, a water level difference of 3 cm was specified in CMS-Flow to simulate the weak current field for Runs 5-8 as close as possible to the measured currents. A water level difference of 5 cm was specified in CMS-Flow to generate the stronger current field pertaining to Runs 9-12. Figure 7 shows the calculated ebb current fields in the steady-state condition (Lin et al. 2008). Figures 8 and 9 show the measured and calculated current speeds along the inlet channel centerline. All four different wave breaking formulas in CMS-Wave were tested. The actual wave data collected in front of the wave generator were analyzed and used as the incident wave input. Other details of CMS-Wave simulations for this laboratory experiments are described elsewhere (Demirbilek et al. 2008; Lin et al. 2008).

Results and Discussion: Measured and calculated wave heights along the entrance channel centerline are compared in Figures 10 to 12 for zero current, moderate ebb current and strong ebb current, respectively. The statistics calculated for Runs 1-4, 5-8, and 9-12, respectively, are presented in Tables 3 to 5. Calculated wave heights and data agree well for zero current (Runs 1-4) for all four wave breaking formulas. The Extended Goda formula agrees slightly better with the measurements than the other

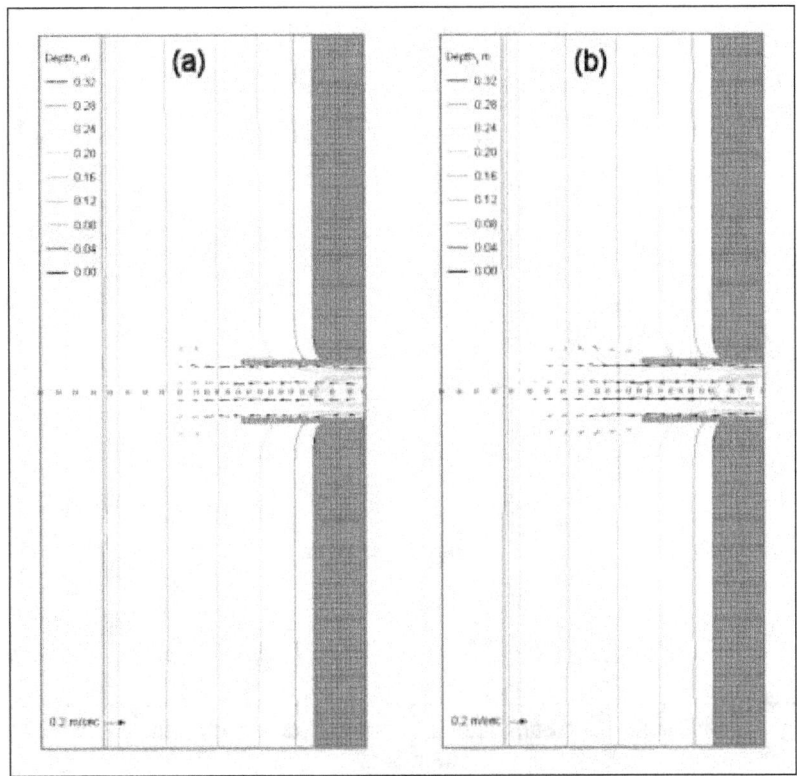

Figure 7. Input current fields for (a) Runs 5-8 with ebb speed ~ 11 cm/sec, and (b) Runs 9-12 with ebb speed ~ 22 cm/sec, and save stations (squares).

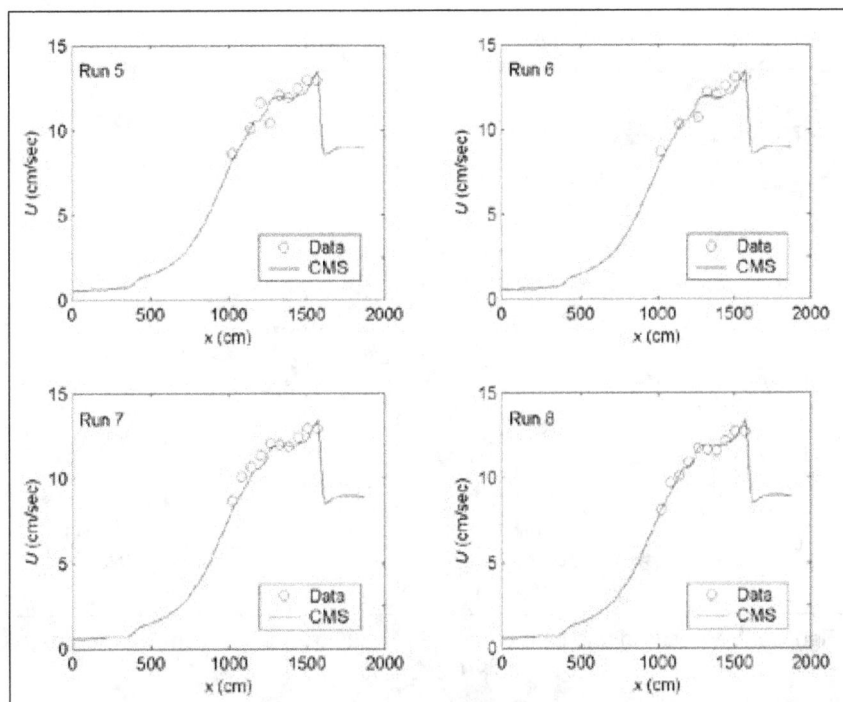

Figure 8. Measured and calculated current speeds along channel centerline, Runs 5-8.

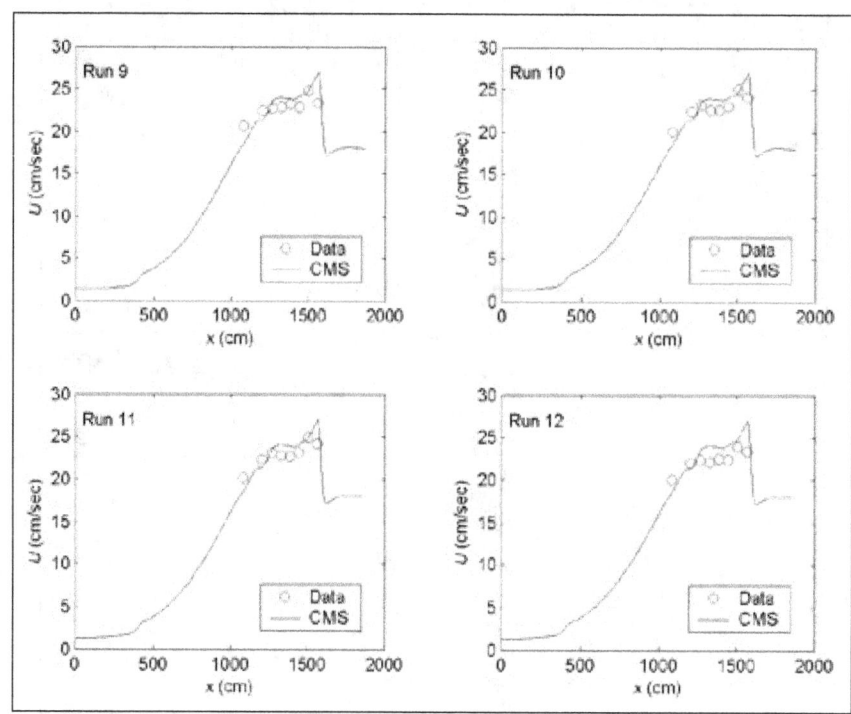

Figure 9. Measured and calculated current speeds along channel centerline, Runs 9-12.

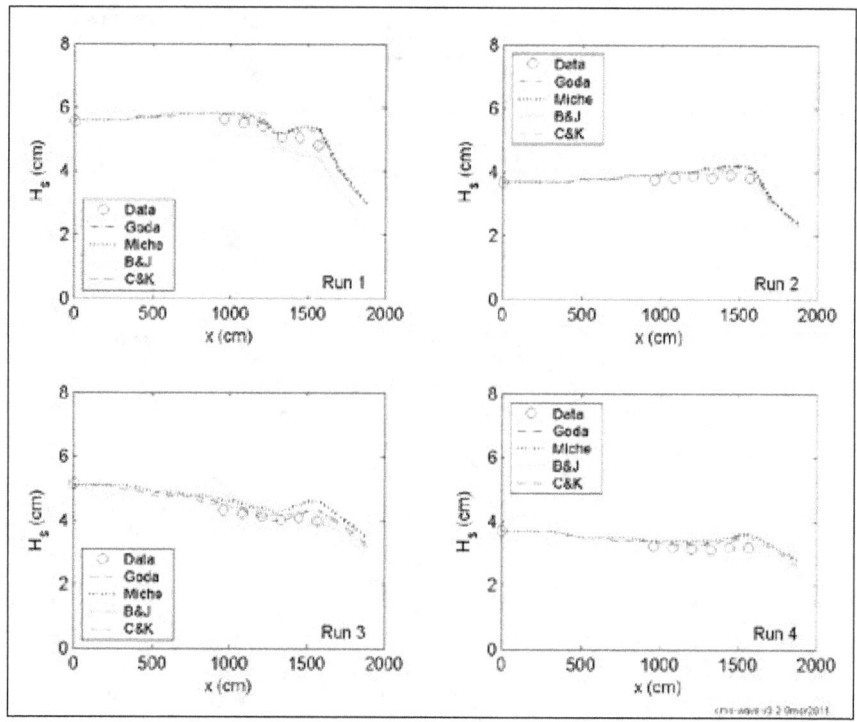

Figure 10. Measured and calculated wave heights along channel centerline, Runs 1-4.

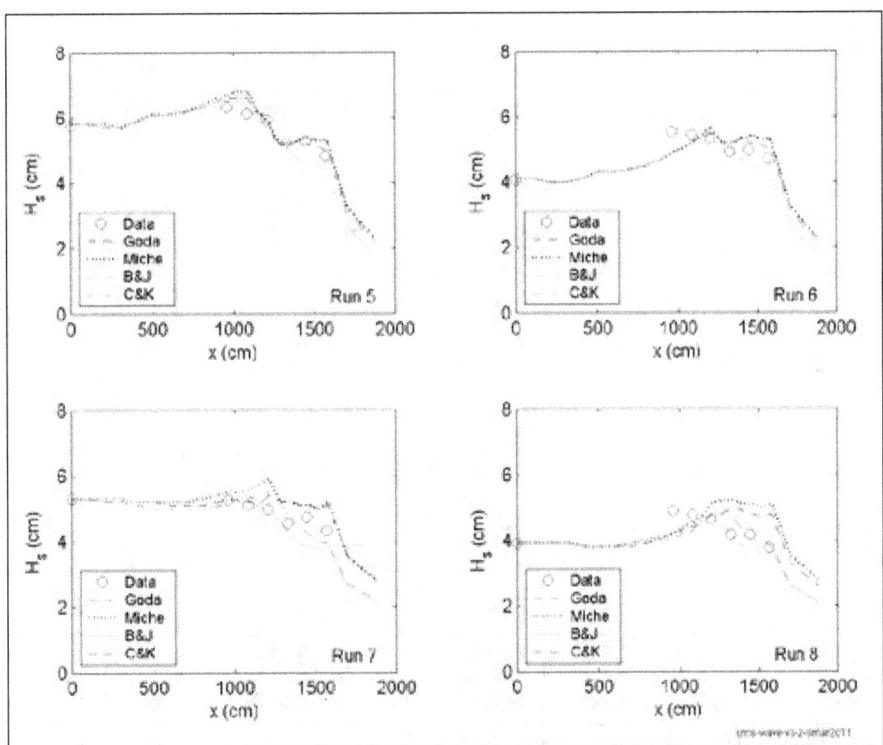

Figure 11. Measured and calculated wave heights along channel centerline, Runs 5-8.

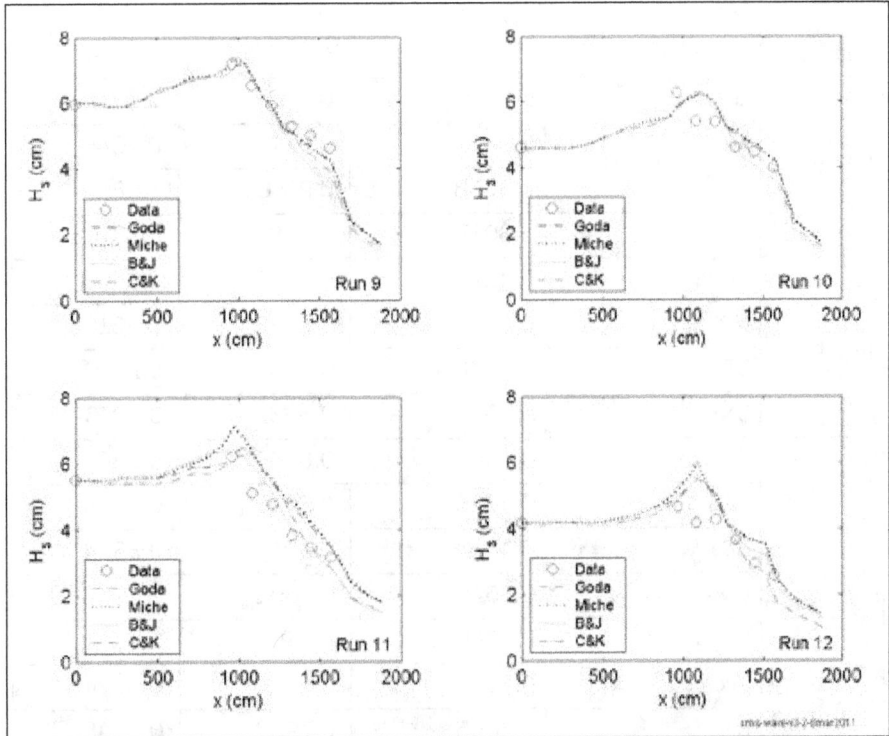

Figure 12. Measured and calculated wave heights along channel centerline, Runs 9-12.

Table 3. Wave height statistics for CHL Inlet experiment Runs 1 to 4.

Case ID	Breaking Formula	RMSE (m)	R	MAE (m)
Run 1	Goda	0.2614	0.8808	0.2067
	Miche	0.2792	0.8815	0.2336
	B&J	0.2826	0.9807	0.2307
	C&K	0.2676	0.9012	0.2324
Run 2	Goda	0.2107	0.8952	0.1851
	Miche	0.2292	0.8365	0.1971
	B&J	0.1267	0.8799	0.1137
	C&K	0.2024	0.8973	0.1785
Run 3	Goda	0.1441	0.9550	0.1013
	Miche	0.3468	0.9315	0.3054
	B&J	0.2328	0.8760	0.1802
	C&K	0.1827	0.9842	0.1600
Run 4	Goda	0.2402	0.8042	0.2111
	Miche	0.2585	0.8067	0.2300
	B&J	0.2520	0.8359	0.2250
	C&K	0.1796	0.9135	0.1586

Table 4. Wave height statistics for CHL Inlet experiment Runs 5 to 8.

Case ID	Breaking Formula	RMSE (m)	R	MAE (m)
Run 5	Goda	0.2739	0.9287	0.1917
	Miche	0.3423	0.9110	0.2346
	B&J	0.3258	0.9746	0.2600
	C&K	0.3017	0.9463	0.1876
Run 6	Goda	0.4112	0.6073	0.3468
	Miche	0.4192	0.6093	0.3610
	B&J	0.3546	0.8553	0.2766
	C&K	0.3587	0.6982	0.3076
Run 7	Goda	0.4406	0.6967	0.3362
	Miche	0.5925	0.5023	0.5049
	B&J	0.4471	0.9044	0.3384
	C&K	0.3541	0.7891	0.2959
Run 8	Goda	0.6266	-0.1913	0.5271
	Miche	0.7968	-0.2355	0.6724
	B&J	0.3055	0.7938	0.2048
	C&K	0.4288	0.3987	0.3178

Table 5. Wave height statistics for CHL Inlet experiment Runs 9 to 12.

Case ID	Breaking Formula	RMSE (m)	R	MAE (m)
Run 9	Goda	0.2240	0.9878	0.1765
	Miche	0.2390	0.9933	0.1805
	B&J	0.5466	0.9792	0.4192
	C&K	0.3709	0.9925	0.2790
Run 10	Goda	0.4450	0.8701	0.3850
	Miche	0.4478	0.8688	0.3867
	B&J	0.3816	0.8979	0.3290
	C&K	0.4526	0.8540	0.3403
Run 11	Goda	0.6855	0.9183	0.5419
	Miche	0.7899	0.9413	0.6905
	B&J	0.3703	0.9728	0.2773
	C&K	0.5722	0.9153	0.4154
Run 12	Goda	0.6472	0.8892	0.4915
	Miche	0.8058	0.8549	0.5860
	B&J	0.6445	0.9159	0.4676
	C&K	0.6114	0.9305	0.4218

three formulas; this is especially true for larger incident wave height (Runs 1 and 3) for the zero current condition. In terms of error metrics, slightly lower RMSE and MAE values for each run correspond to a higher correlation between the model and data.

For Runs 5-12 with an ebb current, the calculated wave height overall agrees better with the measurements for the incident wave of longer period (1.41 sec). This observation holds true regardless of whether there was a large or small wave height and current magnitude, as shown in the upper panel in Figures 11 and 12 (Runs 5, 6, 9, and 10). For the shorter wave period (0.71 sec), the breaking formula of Battjes and Janssen (1978) and the Extended Miche formula (Battjes 1972; Mase et al. 2005b) tended to overestimate the wave height, whereas the formula by Chawla and Kirby (2002) and the Extended Goda formula (Sakai et al. 1989) yielded comparison to data for wave heights. The overall performance of these breaking formulas with wave-current interaction in this experiment revealed that wave height estimates based on the Battjes and Janssen formula were consistently better than estimates obtained with the other formulas.

In the case without current, good model-data comparison for wave heights was obtained with all four wave breaking formulas in CMS-Wave. The MAE is essentially the same for these formulas. In these tests, the Extended Goda formula produced the smallest error (MAE and RMSE), while the Miche's formula resulted in the largest error. Runs 1 and 3 with larger incident wave heights had the largest errors.

For the longer incident wave period with ebb current, good agreement was obtained both for large and small wave heights and current magnitudes with all four breaking formulations. For the shorter wave period, both Battjes and Janssen and the Extended Miche formulas overestimated the wave height, while the Chawla and Kirby and the Extended Goda formulas yielded similar estimates of wave height. Overall, for these wave-current interaction tests, the Battjes and Janssen formula performed consistently about 10 percent better on average than all other wave breaking formulas. Therefore, the Battjes and Janssen wave breaking formula is recommended for wave-current interaction problems at inlet applications, and also for no current conditions. The Extended Goda formula may also be used in applications with no currents.

3.2.2 Test C2-Ex2: Wave breaking experiments on a planar beach

Description: To generate a longshore current with monochromatic incident waves breaking on a planar beach, Visser (1991) conducted eight laboratory experiments, labeled as Run 1 to Run 8. Wave, current, and water level data were collected for a number of incident wave conditions tested on two beach slopes (1:10 and 1:20) and for two different bottom roughnesses. Although these experiments used monochromatic waves (e.g., no irregular wave tests were performed), the data provided are fundamental in checking wave and flow models for wave refraction, shoaling and breaking and wave-induced currents. Lin et al. (2011, 2008) provide additional analysis of these experiments with wave and flow models. Here, only Runs 4 to 7 were selected for model validation because these tests had the same bottom composite slopes and the most complete set of measurements. The beach had a 1:10 slope for the first seaward 1-m distance, 1:20 slope for the next 5-m distance, and a flat bottom for the next 5.9 m to the wave generator. Runs 4 through 6 were conducted on a concrete bed, where the bottom friction is expected to be small and, therefore, neglected in the numerical wave simulation. For Run 7, the 1:20 slope bottom was roughened by a thin layer (0.5-1.0 cm) of gravel grouted on the concrete floor. Table 6 presents the incident wave conditions.

Table 6. Incident wave conditions.

Run	H_s (cm)*	T_p (sec)*	f_p (Hz)**	θ (deg)***
4	7.8	1.02	0.98	15.4
5	7.1	1.85	0.54	15.4
6	6.9	0.70	1.43	15.4
7	7.8	1.02	0.98	15.4

* Monochromatic wave.
** $f_p = 1/T_p$.
*** Wave direction relative to shore-normal.

Model setup and parameters: CMS-Wave was run at laboratory scale, and unless otherwise noted, simulations were made with the default parameters (see Appendix A). The model grid consisted of 90 cross-shore and 243 alongshore square cells, each 10 cm × 10 cm, to cover the entire basin in these experiments. The spectral transformation was computed in CMS-Wave on 11 frequency bins (covering the range of +/-0.05 Hz of the incident monochromatic wave frequency at 0.01-Hz increment) and 35 direction bins (covering a half-plane with 5-deg spacing). The incident monochromatic, unidirectional wave spectrum was specified in a single frequency and direction bin at the seaward boundary. The input current field was interpolated cross-shore and averaged alongshore from the data. For Run 7 with the gravel floor, a constant Darcy-Weisbach type bottom friction coefficient of 0.01 was specified in CMS-Wave (Lin et al. 2008). This value was determined by a trial-and-error run and was used in all other runs. The model was not calibrated with data.

Results and Discussion: Figure 13 shows an example of input current and water level fields derived from the measurements used in the CMS-Wave simulation for Run 4. "Exp" on these figures refers to the "Run" number. Figure 14 shows a comparison of the measured and calculated cross-shore wave heights for Runs 4 through 7. The calculated wave height agrees well with the measurements for four different depth-limiting breaking formulas implemented in CMS-Wave. Table 7 presents the statistics calculated for Runs 4 to 7. Comparatively smaller MAE values occur for Runs 6 and 7 using the Extended Goda as well as the Battjes and Jassen wave breaking formula; these error values are nearly the same for both formulas.

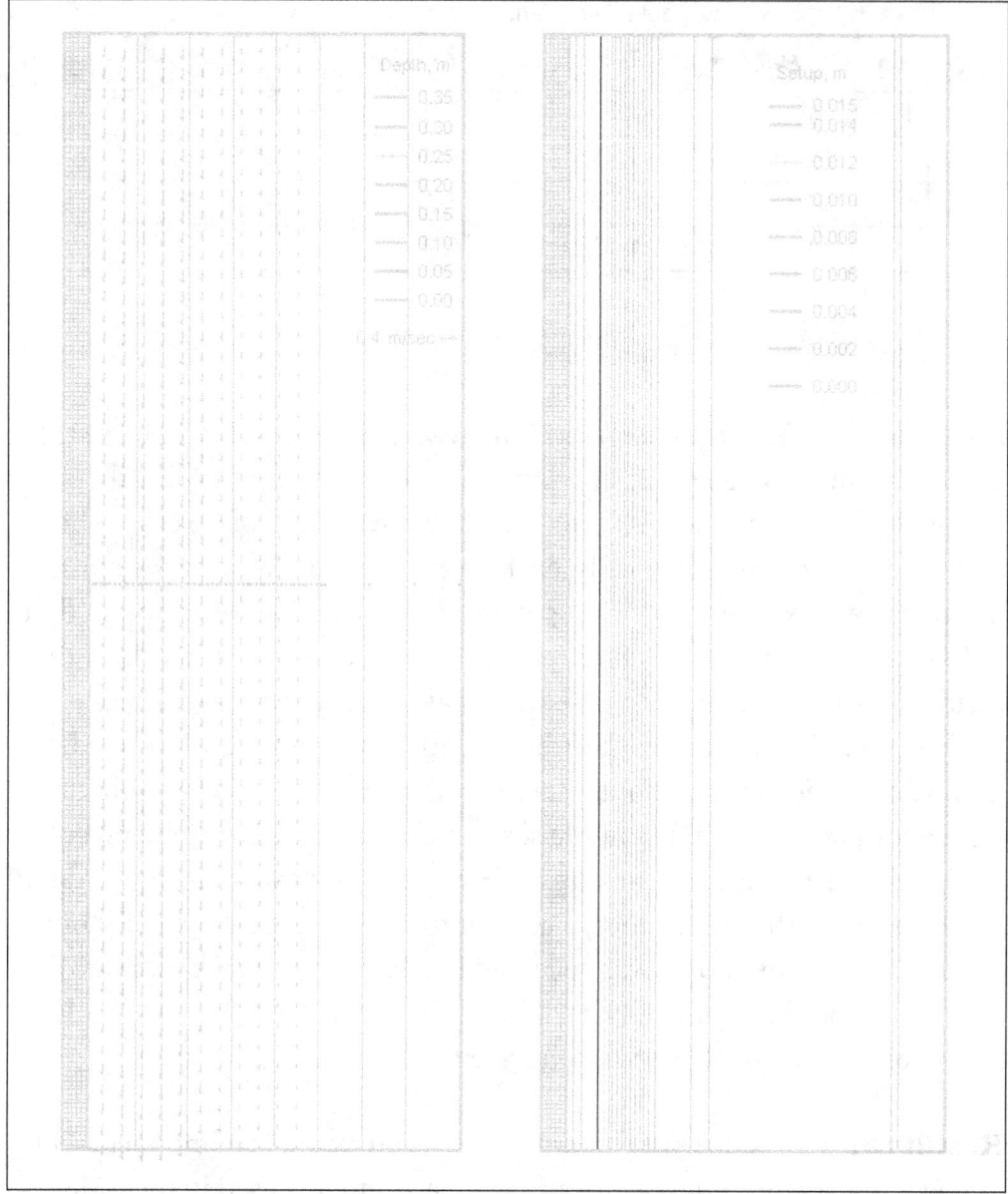

Figure 13. Input current and wave setup fields with save stations (dot) for Run 4.

For oblique monochromatic waves breaking on a planar beach and interacting with the longshore current, the calculated wave heights agree with data for all four wave breaking formulas in CMS-Wave. Although these results are for monochromatic waves, other test cases provided under Categories 2 and 3 include applications with irregular waves. Therefore, this test case was necessary for validation of wave model, to show that CMS-Wave represents wave refraction, shoaling and breaking and wave-induced currents accurately, and that the largest error in the calculated wave height was less than 5 percent of the incident wave height along a planar beach.

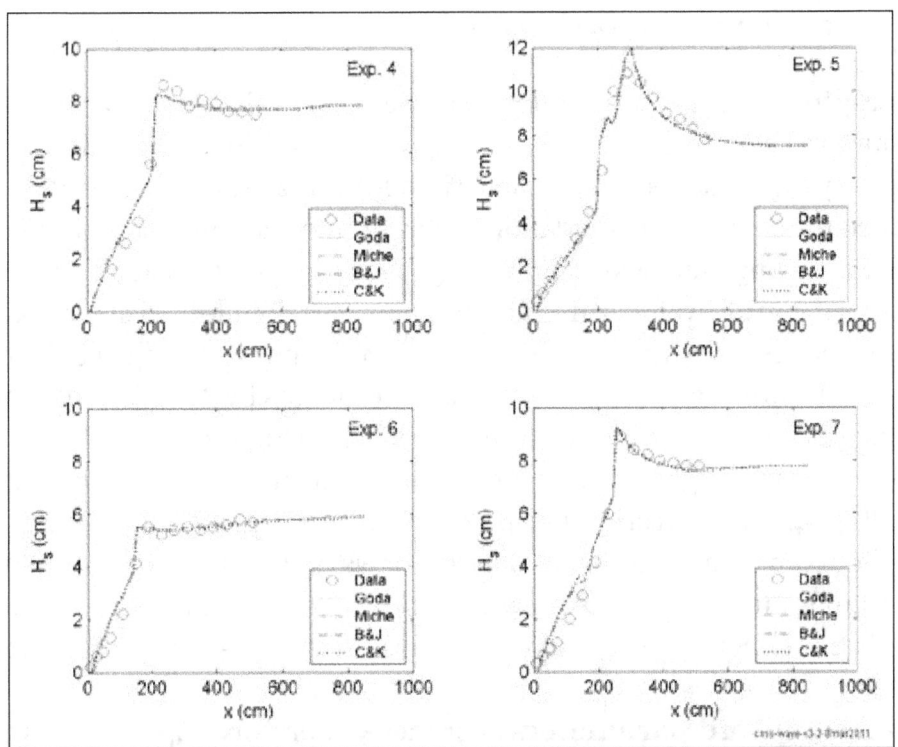

Figure 14. Measured and calculated wave heights, Runs 4-7.

Table 7. Wave height statistics for planar beach Runs 4 to 7.

Case ID	Breaking Formula	RMSE (m)	R	MAE (m)
Run 4	Goda	0.4124	0.9931	0.3433
	Miche	0.4124	0.9931	0.3433
	B&J	0.4124	0.9931	0.3433
	C&K	0.4124	0.9931	0.3433
Run 5	Goda	0.6413	0.9865	0.4840
	Miche	0.6089	0.9865	0.4147
	B&J	0.6089	0.9865	0.4147
	C&K	0.6118	0.9864	0.4193
Run 6	Goda	0.3445	0.9916	0.2267
	Miche	0.3445	0.9916	0.2267
	B&J	0.3445	0.9916	0.2267
	C&K	0.3479	0.9904	0.2400
Run 7	Goda	0.3617	0.9953	0.2567
	Miche	0.4509	0.9938	0.3333
	B&J	0.4440	0.9940	0.3300
	C&K	0.4715	0.9931	0.3600

3.2.3   Test C2-Ex3: Wave runup on impermeable uniform slope

Description: The purpose of this test case was to validate the wave runup on a uniform slope calculated by CMS-Wave with two datasets. Two laboratory experiments (Ahrens and Titus 1981; Mase and Iwagaki 1984), supplied the data for this validation. Detailed information on measurements, including dimensions of flume, gauge types and data analyses performed, are all available from these references. Random incident waves were generated in both experiments in a wave flume consisting of a flat bottom offshore of a sloping beach. The experiments by Ahrens and Titus included 275 wave conditions (Ahrens and Heimbaugh 1988), with the significant wave heights ranging from 4 to 20 cm and spectral peak periods from 1.1 to 4.5 sec, and six uniform slopes (1:1, 2:3, 1:2, 2:5, 1:3, 1:4). The experiments by Mase and Iwagaki used 120 wave conditions (Mase 1989), with significant wave heights ranging from 2.7 to 11 cm and spectral peak periods from 0.8 to 2.5 sec, and four uniform slopes (1:5, 1:10, 1:20, 1:30).

Model setup and parameters: For these laboratory applications, CMS-Wave was set up for the laboratory scale, and unless noted otherwise, all numerical simulations were performed with the default parameters listed in Appendix A. The model grid consisted of 1,750 cross-shore and 100 alongshore square cells, each 2 cm × 2 cm, covering a 10-m long flat bottom and a 25-m long slope sections. The spectral wave transformation was computed with 30 frequency bins (from 0.1 to 1.26 Hz at 0.04-Hz increment for the Ahrens and Titus experiments, and from 0.02 to 2.05 Hz at 0.07-Hz increment for the Mase and Iwagaki experiments), and 35 direction bins (covering a half-plane with 5-deg spacing). Bottom friction and wave reflection were neglected in these simulations. The default parameters were used including the Extended Goda wave breaking formula. The runup calculation was described in Lin et al. (2011).

Results and Discussion: Figures 15 and 16 show the calculated and measured 2% exceedence wave runup ($R_{2\%}$) for all experiments (395) conducted by Ahrens and Titus (1981) and by Mase and Iwagaki (1984). The 45-deg line in the figures represents a perfect match of calculated and measured values. Tables 8 and 9 present the statistics between the calculated and measured $R_{2\%}$ values for these two experimental studies. The calculated wave runup values correlate with data differently depending on slope. Overall, the MAE of calculated runup was small for all test cases except for the steepest slope (1:1), for which CMS-Wave overestimated the runup (open-circles in Figure 15). A higher correlation, greater than

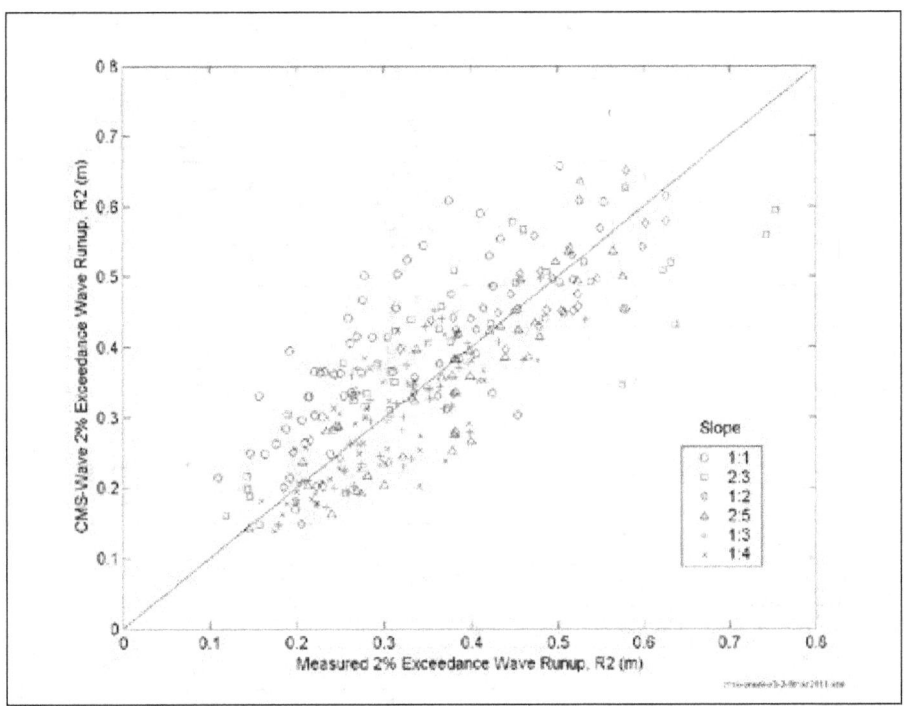

Figure 15. Comparison of 2% exceedence wave runup for slopes of 1:1 to 1:4.

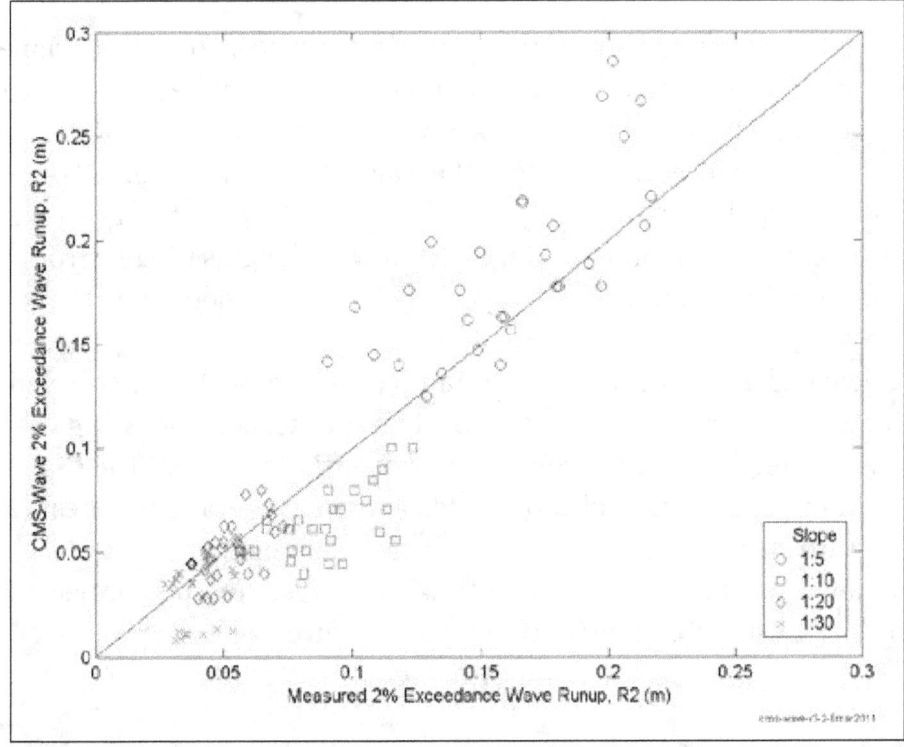

Figure 16. Comparison of 2% exceedence wave runup for slopes of 1:5 to 1:30.

Table 8. Statistics of measured and calculated wave runup on slopes of 1:1 to 1:4.

Slope	RMSE (m)	R	MAE (m)
1:1	0.1443	0.7209	0.1270
2:3	0.1013	0.8059	0.0842
1:2	0.0626	0.8779	0.0506
2:5	0.0550	0.8790	0.0471
1:3	0.0467	0.8396	0.0361
1:4	0.0627	0.7276	0.0458
All	0.0864	0.7827	0.0655

Table 9. Statistics of measured and calculated wave runup for slopes of 1:5 to 1:30.

Slope	RMSE (m)	R	MAE (m)
1:5	0.0385	0.7143	0.0292
1:10	0.0299	0.6217	0.0259
1:20	0.0117	0.5737	0.0098
1:30	0.0164	0.4163	0.0124
All	0.0264	0.9256	0.0193

85 percent was obtained for gentler slopes (1:5 to 1:30). For steeper slopes, more wave reflection (not calculated in these simulations because reflected waves were removed in data analyses) and less wave energy loss takes place on the slope section, and these affect the wave runup calculations. The CMS-Wave calculations of $R_{2\%}$ agree with data (70 to 90 percent correlation, with errors less than 15 cm for incident wave heights ranging from 3 cm to 20 cm) for all test slopes, especially for gentler slopes (1:5 to 1:30).

The calculated 2% exceedence wave runup ($R_{2\%}$) shows better correlation with data for flatter slopes (1:5 to 1:30) than the steeper slopes (1:5 to 1:1). This is expected because the runup is less over flatter slopes. The CMS-Wave runup function is applicable to coastal structures and beaches with the seaward slope less than 1:5. This test case demonstrates that CMS-Wave can be used for preliminary estimates of wave runup in projects. For steep slopes, estimates of wave runup may require using the phase-resolving nonlinear models or physical modeling studies.

3.2.4 Test C2-Ex4: Experiments for Cleveland Harbor, Ohio

Description: The purpose of this test was two-fold: a) compare CMS-Wave calculations to data from a physical modeling study that investigated wave propagation at the entrance of Cleveland Harbor (data were in prototype scale); and b) inform users about a recent CMS-Wave model verification and validation study by Demirbilek et al. (2010) that provided wave estimates in and around the Cleveland Harbor complex in support of planned harbor modification works. A 1:100-scale physical model of Cleveland Harbor, Ohio, was constructed in 1980-1981 at the Waterways Experiment Station (WES) to investigate the effects of waves, currents, and river flow on ship maneuverability (Bottin 1983) in the entrance and within the harbor complex.

Cleveland Harbor is situated on the south shore of Lake Erie, and is protected by two breakwaters with a combined length over 6 miles (10 km). The east breakwater consists of rubble mound stone while the west breakwater is mainly composed of concrete caissons. The Harbor has two entrances: the west (main) entrance is located lakeward of the Cuyahoga River Mouth and the east entrance is at the eastern end of the east breakwater (Figure 17). In a recent study by Demirbilek et al. (2010), incident waves were transformed from the Wave Information Study (WIS) Station 10 located offshore of the harbor to the project site (Figure 17) using CMS-Wave. This earlier study described details of the numerical modeling, including a study plan, tasks, modeling approach and estimates of wave parameters developed for different project needs. Results by Demirbilek at al. (2010) are not duplicated here, and interested readers should refer to their repot.

The laboratory experiment tested 126 cases, consisting of 20 incident wave heights, 12 wave periods, 3 wave directions, 3 lake water levels, and two river discharges. Table 10 presents the range of physical model test conditions converted to prototype using a scale of 1:100 in order to compare to prototype results reported both by Bottin (1983). Figure 18 shows the harbor main entrance and 29 wave gauge locations in the physical model.

Model setup and parameters: The CMS grid was oriented East-West, with the offshore boundary at the 16 m depth contour, and extended from the most westward to the furthest eastward ends of the Cleveland Harbor complex (Figure 19). CMS-Wave was run in the prototype because the laboratory data were reported in the prototype scale. The computational domain covered approximately 172 square kilometers. It consisted of 860 × 2000 cells with a uniform cell size of 10 m x 10 m. CMS-Wave was run for four selected cases that represented the most probable wave with normal

Figure 17. Location of Cleveland Harbor, OH and WIS Station 10.

Table 10. Cleveland Harbor laboratory experiment condition (in prototype units).

Test Conditions	Total No. of Conditions	Range (or Value)
Wave Height (m)	20	1.2 to 4.2
Wave Period (sec)	12	6 to 10
Wave Direction (deg)	3	279, 326, 17
Lake Water Level (m)	3	0, 0.8, 1.5
River Discharge (m³/sec)	2	22.6 and 226

and extreme water levels and river discharges. Both monochromatic and unidirectional incident waves were used in the laboratory experiments. Because no data were available on the background current field used in the laboratory experiments, the input to CMS-Wave was pre-calculated using CMS-Flow. Table 11 presents the four cases evaluated. The wave transformation used a spectral grid of 30 frequency bins (0.04 to 0.33 Hz with 0.01-Hz increment) and 35 direction bins (covering a half-plane with 5-deg spacing). Bottom friction and wave reflection were included in the CMS simulations based on model-data calibration tests. Constant values of Manning ($n = 0.025$) (default) and the wave reflection coefficient (= 0.5), were used. The infra-gravity wave capability and wave transmission and overtopping (of breakwaters) features were all triggered in these CMS-Wave simulations. The default parameters were used for other parameters with the Extended Goda wave breaking formula and a maximum diffraction intensity of $\kappa = 4$.

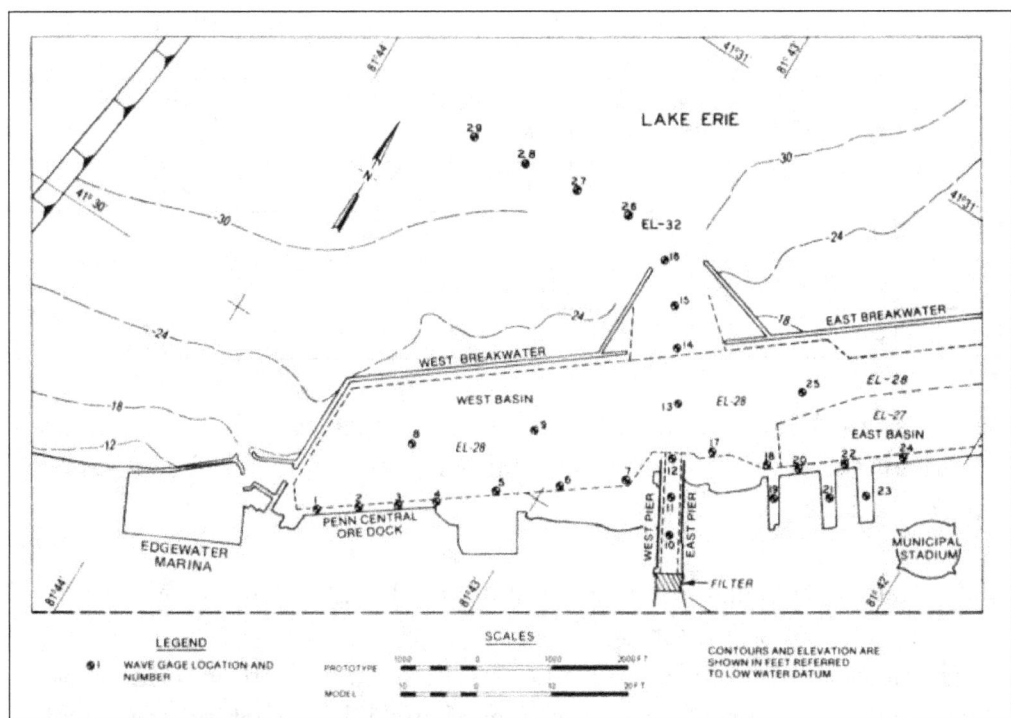

Figure 18. Physical model of Cleveland Harbor main entrance and 29 wave gauge locations
(from Bottin 1983).

Figure 19. CMS-Wave modeling domain for Cleveland Harbor.

Table 11. Cleveland Harbor wave, water level, and river discharge test cases (in prototype units).

Case ID	Wave Height (m)	Wave Period (sec)	Wave Direction (deg)	Water Level (m)	River Discharge (m³/sec)
1	3.14	8	326	0	22.7
2	3.14	8	326	0	227
3	3.14	8	326	1.46	22.7
4	3.14	8	326	1.46	227

Results and Discussion: Demirbilek et al. (2010) verified and validated the calculations of CMS-Wave using three methods and a different set of wave conditions:

a. Numerical modeling results were compared to data from one of the test cases in the laboratory study;
b. The numerically predicted wave heights through the harbor entrance were verified by comparing to estimates based on analytical solutions available in the Coastal Engineering Manual (CEM 2006), and
c. Transmitted waves over the East Breakwater were verified by comparing to estimates obtained using empirical formulas from the CEM.

In these comparisons, agreement between model results and data and with analytical and empirical methods varied depending on incident wave conditions. Greater differences occurred for storm waves at high lake water levels. These results will not be repeated here and interested readers may refer to Demirbilek et al. (2010) for details of the analyses performed.

For the simulations listed in Table 11, Figures 20 to 23 show the calculated and measured wave heights for Cases 1 to 4, respectively. The 45-deg line shown in the figures indicates a perfect match between calculated and measured wave heights. Table 12 presents the statistics between the calculated and measured wave heights. The calculated wave heights correlate well with data for the two water level and river discharge conditions tested. Good correlation does not necessarily mean error statistics are small. Comparatively higher errors (0.2 to 0.5 m range) occurred for comparisons in the sheltered regions affected by wave transmission, and also at locations affected by wave diffraction and reflection processes. These errors are partly due to differences between numerical model inputs and laboratory setup. As examples, the harbor bathymetry and structures data used in the physical model study were not available. Wave parameters

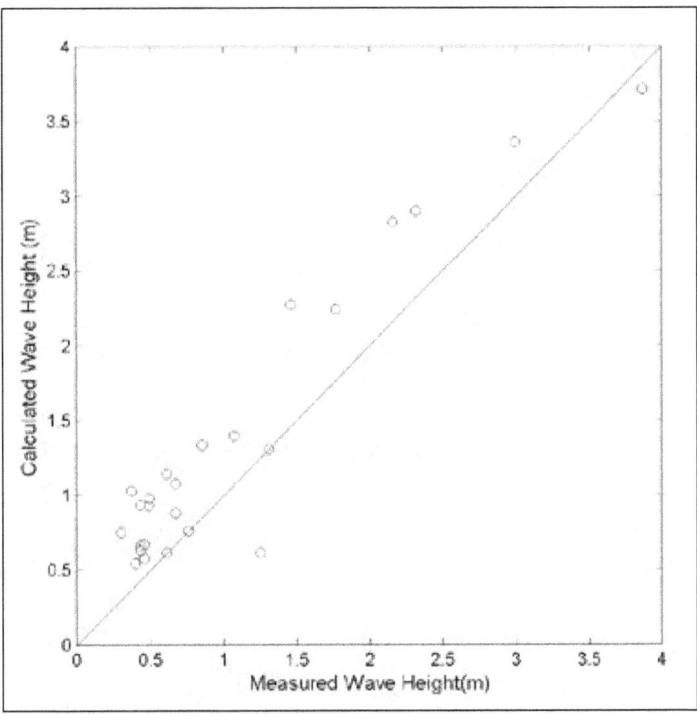

Figure 20. Comparison of Cleveland Harbor test Case 1 wave heights.

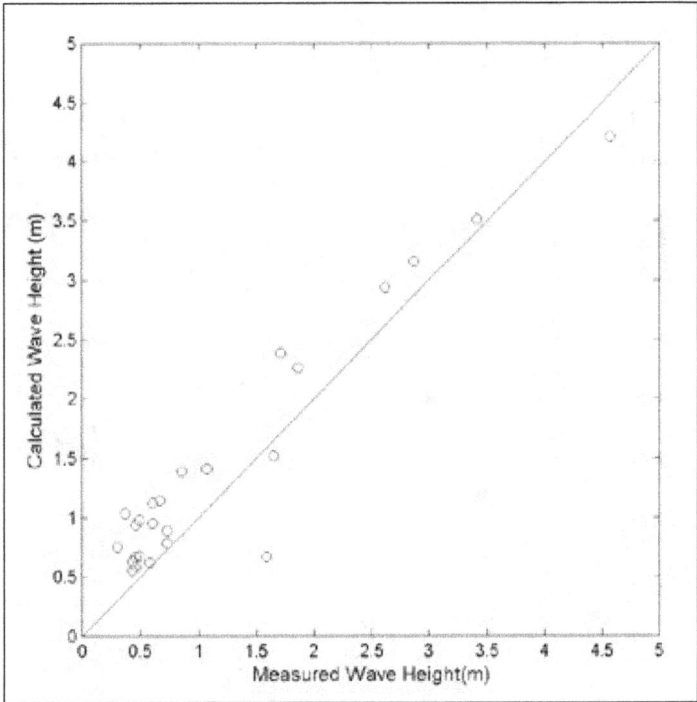

Figure 21. Comparison of Cleveland Harbor test Case 2 wave heights.

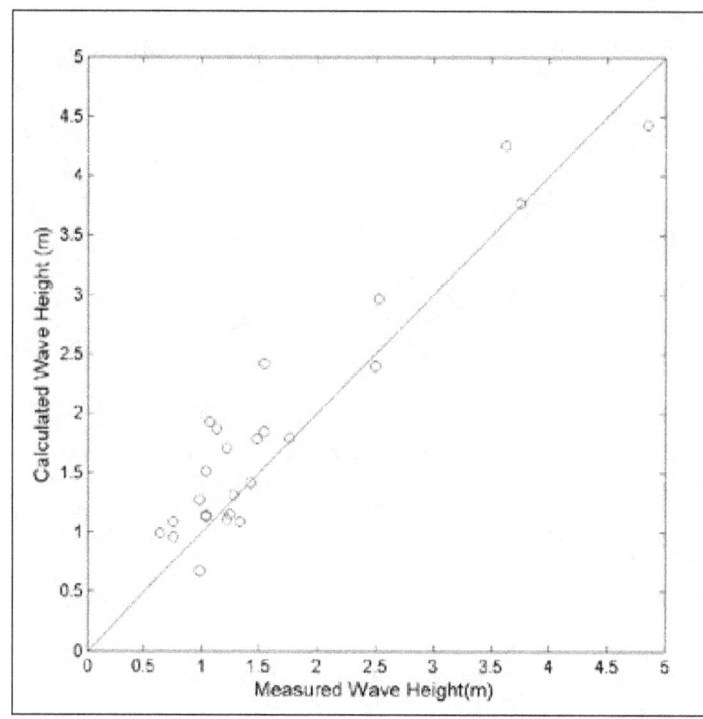

Figure 22. Comparison of Cleveland Harbor test Case 3 wave
heights.

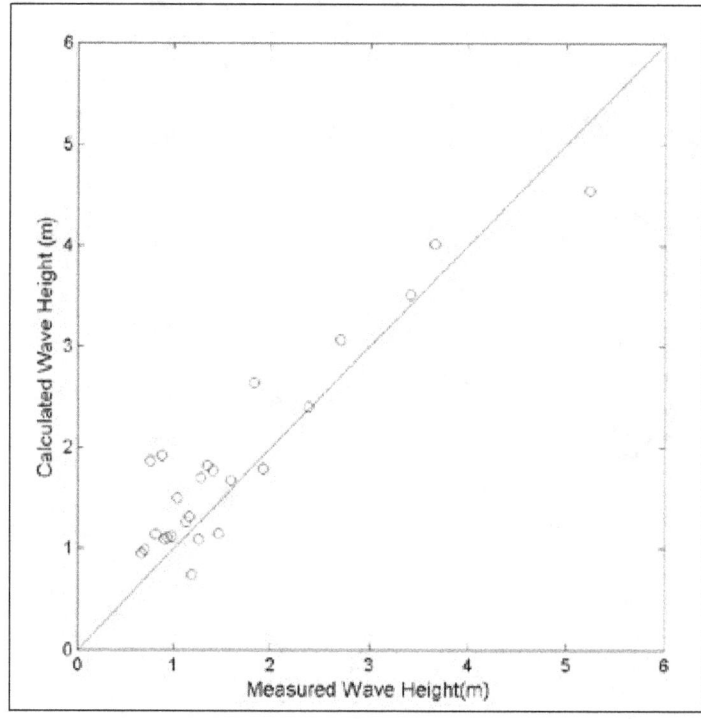

Figure 23. Comparison of Cleveland Harbor test Case 4 wave
heights.

Table 12. Cleveland Harbor measured and calculated wave
height comparison statistics.

Case ID	RMSE (m)	R	MAE (m)
1	0.43	0.94	0.36
2	0.41	0.95	0.34
3	0.40	0.94	0.31
4	0.46	0.93	0.36

used in the laboratory experiments could have been different than those used in present numerical modeling study and some were not specified in the 1983 study report. The corresponding monochromatic waves generated in the CMS-Wave may not be identical to monochromatic waves used in the experiments. The difference in the bathymetries and input spectra between the laboratory and numerical models is one of the causes of the observed discrepancy between model results and data. Secondly, the attenuation of monochromatic waves is generally weaker in numerical modeling (Lin et al. 2008), that is regular waves attenuate less as compared to irregular waves with similar integral wave parameters (height, period, direction).

Wave heights calculated with CMS-Wave compared reasonably well to data from the Cleveland Harbor physical model, with some large errors, as much as 50% in some cases, for certain wave conditions and gauge locations. The best agreement was with the higher waves outside the entrance to the harbor. These discrepancies are expected as explained above, and also because numerical model simulations considered the combined effects of wave diffraction, reflection, transmission and overtopping at breakwaters. Inclusion of bottom friction, background current fields and infra-gravity effects had minimal effect on model-data comparison, suggesting that in this test case, the dominant wave processes were wave diffraction, reflection, wave transmission/overtopping of the structures, entrance losses (not represented in CMS-Wave), and wave-current interaction. The calculated waves inside the harbor complex were generally slightly overpredicted, suggesting that there was too much wave transmission or overtopping considered in the simulations. No attempt was made to calibrate the model to data, and model was run with default parameters in order to objectively evaluate its suitability for these types of applications.

Overall, in spite of numerous differences in the bathymetry and changes in the harbor geometry and structures and input conditions between numerical and physical model studies, the predictions and data exhibit similar trends, and the overall model-data comparison was satisfactory. Quantitatively, the results were similar, but wave heights predicted by CMS-Wave decayed through the entrance faster than waves in the laboratory study. The comparison also showed that waves near the piers at the Cuyahoga River mouth outside the entrance were similar to the CMS-Wave results. Comparison of model-CEM diffraction diagrams for a gap problem was also performed to further evaluate the model's diffraction estimates. This comparison is omitted here since model results were verified for a gap problem in Chapter 1 and interested readers can find details in Demirbilek et al. (2010).

For harbor applications, CMS-Wave simulations should include all important mechanisms such as wave diffraction, reflection, transmission, overtopping of breakwaters, wave-current interaction, and infra-gravity wave effects. It is not possible to isolate the individual importance of each of these processes. Therefore, this test case validates nearshore wave heights calculated by CMS-Wave at the Cleveland Harbor entrance, and indicates that the model is suitable for these types of engineering applications. Granted, this is an "extreme" application for a spectral wave model because strong reflection, diffraction, runup/overtopping and wave-current interactions are challenging in this class of wave models. However, estimates given here are appropriate for planning and feasibility level studies, and final design estimates should be checked with phase-resolving wave models (e.g., CGWAVE or BOUSS-2D), which need the transformed wave conditions from CMS-Wave offshore of the harbor complex.

4   Category 3 Test Cases: Field Studies with Data

4.1   Overview

This Chapter includes the Category 3 cases which represent applications of CMS-Wave to field studies containing data. These V&V test cases are listed below in two groups: completed and in progress (under study). The cases under study will be presented in a future companion report.

Completed cases:

1. Matagorda Bay, Texas,
2. Grays Harbor, Washington,
3. Mouth of Columbia River, Oregon/Washington,
4. Southeast Oahu coast, Hawaii,
5. Recent FRF, North Carolina, wave measurements,
6. Mississippi Coastal Improvement Program (MsCIP), and
7. Indian River County, Florida.

Cases in progress:

1. Pillar Point Harbor, California,
2. Noyo Harbor, California, and
3. Galveston Bay, Texas.

The three statistical measures defined in Chapter 3 (see Equations (1), (2) and (3)) are used here in the evaluation of model performance (model-data comparison). For all field test cases investigated, the bottom friction was included in the CMS-Wave simulations. The recommended default value for sandy beds is $C_f = 0.005$ for the Darcy-Weisbach coefficient, and $n = 0.025$ for the equivalent Manning's coefficient.

4.2   Test cases

4.2.1   Test C3-Ex1: Matagorda Bay, Texas

Description: The purpose of this test case was to validate local wind-wave generation and full-plane capabilities of the CMS-Wave for calculating wave heights in an enclosed bay connected to an inter-coastal waterway (ICW) as well as to a major water body (Gulf of Mexico).

Matagorda Bay is located on the central coast of Texas, with a surface area of approximately 930 km² and quite shallow with depths ranging between 2 to 4 m. The tidal prism of the bay is large because of the large bay surface area, despite the modest tidal range of about 0.33 m in the bay. The bay is separated from the Gulf of Mexico by Matagorda Island and Matagorda Peninsula. Freshwater discharge that originates from the Colorado River and the Lavaca River is less than 10 percent of the daily tidal exchange through the two inlets with the Gulf of Mexico. Local wind is the dominant forcing for generating waves in the bay.

Directional wave data and water level data were collected with a bottom-mounted Acoustic Doppler Profiler (ADP) in 3.8 m of water for the time period from September to December 2005, at a middle bay location (Puckette 2006). The coordinates of this wave measurement station, MBWAV, were 28°31.285'N, 96°24.423'W. Local wind and tide data are available from a NOAA Station 87737011 at Port O'Connor (28°26.8'N, 96°23.8'W) in the southwest corner of the bay. Figure 24 shows the wind, tide, and wave data-collection locations. Figure 25 shows the hourly wind, tide, and wave data collected in September-December 2005. Water level data collected at MBWAV and Port O'Connor show that the spatial variation of water level in the bay occurs with passage of a cold front system and strong winds. The strong wind condition on 24 September 2005 was Hurricane Rita.

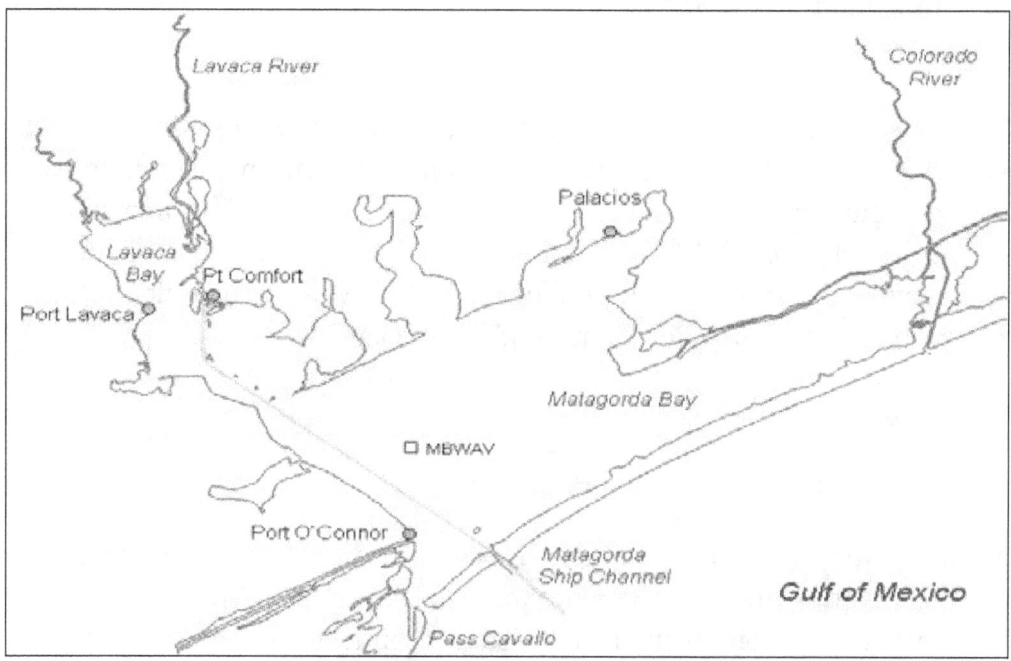

Figure 24. Wind, tides, and wave data-collection stations in Matagorda Bay.

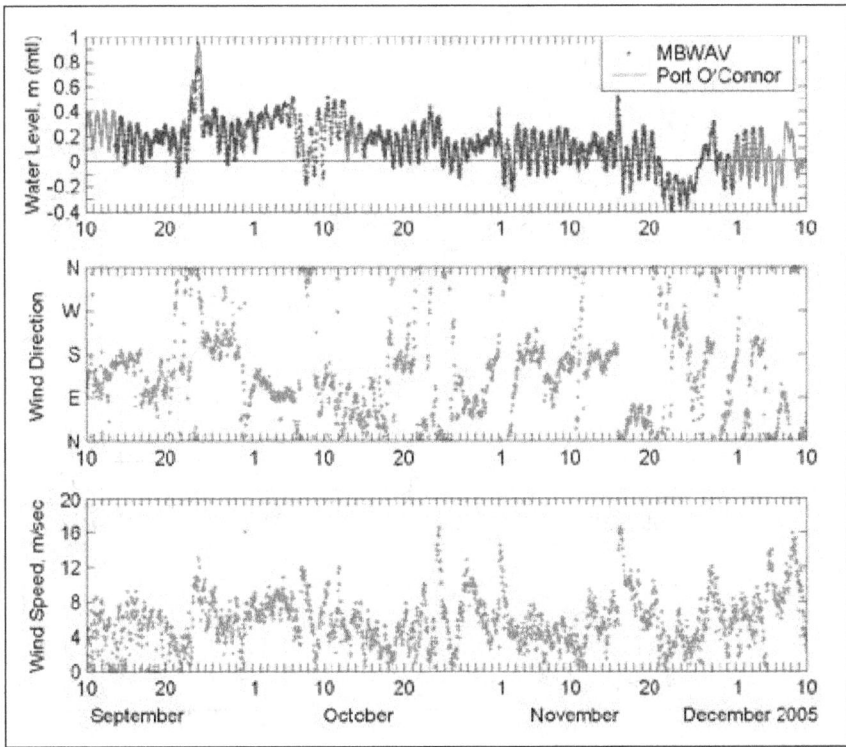

Figure 25. Matagorda Bay wind and water level data, September–December 2005.

Model setup and parameters: CMS-Wave simulations were performed on a rectangular domain covering the entire bay and connecting parts of the ICW and GOM. The default parameters were used (see Appendix A) with the Extended Goda wave breaking formula and a diffraction intensity of κ =4. Wind input was used to drive model boundaries. Wave generation and growth, and interaction of waves with bathymetry and current were of concern. The numerical grid consisted of 153 × 279 cells with variable cell spacing of 29 m to 1,600 m. Wave generation and propagation were computed on a spectral grid of 40 frequency bins (0.06 to 0.45 Hz with 0.01-Hz increment) and 35 direction bins (covering a half-plane with 5-deg spacing and full-plane with 10-deg spacing). Measured wind (Port O'Connor) and tidal elevation (MBWAV) were input to the model, and a constant Manning coefficient of n = 0.025 (default value) was specified for bottom friction.

Results and Discussion: Waves generated by the wind were small to moderate in the middle of the bay during the data collection period. Table 13a lists four large wave events of interest observed in the data that were produced by relatively strong wind, greater than 10 m/sec. Waves in this bay were generally less than 1 m in height and had short periods,

Table 13a. Comparison of measured and calculated waves at MBWAV.

Date	Time (GMT)	Wind Speed (m/sec) / Wind Direction (deg)	Depth, MTL (m)	Measured Wave ht (m) / Period (sec) / Direction (deg)	Calculated Wave ht (m) / Period (sec) / Direction (deg)
9/24/2005	0300	10.5 / 5	3.95	0.62 / 3.2/ 15	0.67 / 2.9 / 10
10/24/2005	0800	15.5 / 10	3.95	0.95 / 4.0/ 25	0.97 / 3.1 / 15
11/1/2005	0200	14.0 / 10	3.85	0.85 / 3.4/ 20	0.88 / 3.1 / 15
11/16/2005	0900	16.5 / 15	3.85	1.02 / 3.6/ 20	1.01 / 3.1 / 20

Table 13b. Statistics of measured and calculated wave heights, periods and directions.

Wave Parameter	Half/Full-Plane	RMSE	R	MAE
Significant height	Half plane	0.10 m	0.70	0.07 m
	Full plane	0.09 m	0.73	0.07 m
Peak Period	Half plane	0.5 sec	0.46	0.32 sec
	Full plane	0.49 sec	0.36	0.31 sec
Mean direction	Half plane	56.7 deg	0.69	41.1 deg
	Full plane	58.3 deg	0.66	42.2 deg

below 5 sec during the measurement period. The largest waves were generated on 24 September 2005 under the strong winds of Hurricane Rita. The other three large wave events occurred on 24 October, 1 November, and 16 November 2005, as produced by cold fronts. These large waves were generated by a wind directed between north and north-northeast.

Measured and calculated significant wave height, spectral peak period, and mean wave direction for the four large wave events simulated are listed in Table 13a. Figure 26 shows an example of the wave field generated for 0700 GMT 24 September 2005. Calculated and measured wave height, period and direction obtained with the full-plane model agreed with data at four measurement locations. The largest difference between model calculations and data was less than 5 cm for the wave heights, 0.9 sec for the wave periods, and 10 deg for the wave directions.

The CMS-Wave validation was conducted for a 22-day simulation from 9 to 30 September 2005 in half-plane and in full-plane mode. Large waves were generated in the Bay on 24 September 2005 under the strong winds resulting from Hurricane Rita. These large waves were generated by a wind directed from north and north-northeast. Figure 27 shows the time series of calculated significant wave height, peak period and mean

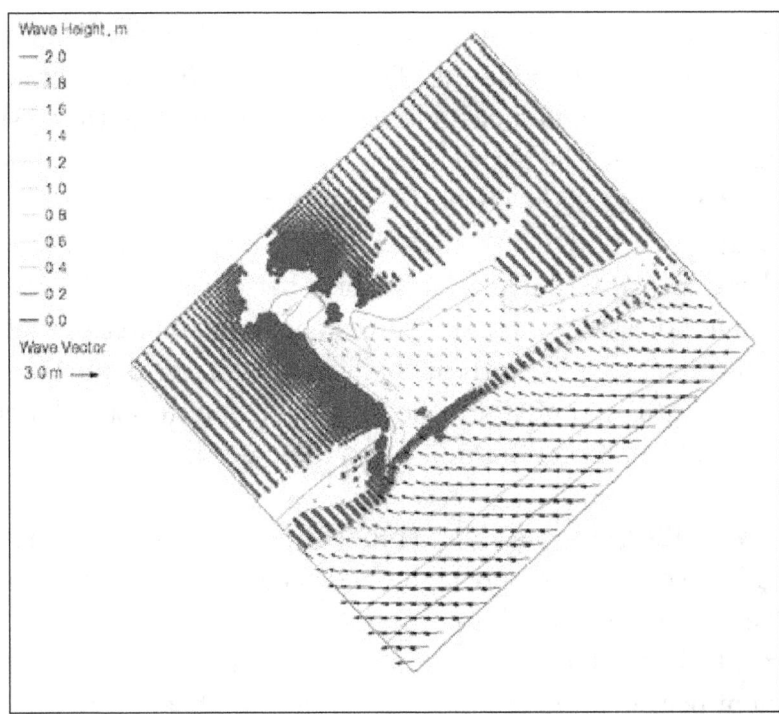

Figure 26. CMS-Wave calculated wave field at 0700 GMT, 24 September 2005.

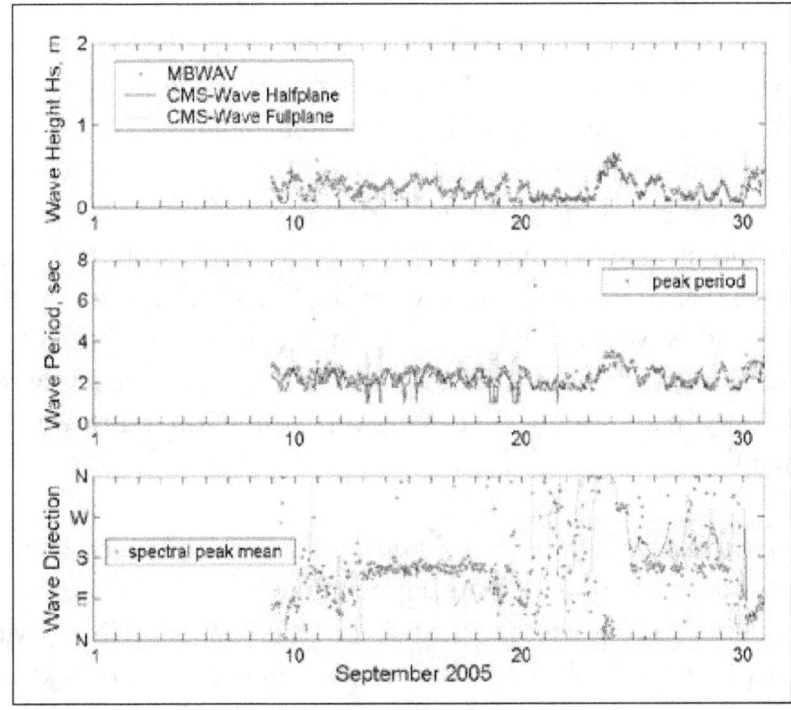

Figure 27. Comparison of wave heights, periods, and directions, 9 to 30 September 2005.

direction with data. The statistics between the calculated and measured wave heights are given in Table 13b. The correlation of 0.7 is obtained between the calculated and measured wave heights with both half-plane and full-plane models. The correlation of calculated and measured periods is relatively low owing to the generally small range of wind wave periods (between 2 and 4 seconds) in the bay. The RMSE and MAE for the calculated and measured wave periods show better results with their values equal to or less than 0.5 sec. The calculated and measured wave directions correlate well for both the half-plane and full-plane cases. However, the RMSE and MAE for the calculated and measured mean wave direction are not small (in the range of 40 to 60 deg).

In summary, CMS-Wave was able to model the generation and propagation of wind-waves for relatively high wind speeds within this shallow bay region. Overall, the calculated wave parameters were similar to the measurements for the four largest storm events during the 3 month measurement period, although there are some significant differences between the measured and the calculated wave direction and period. The calculated spectral peak wave period was slightly underestimated, probably because the nonlinear wave energy transfer is more pronounced in the shallow water than in deep water and would be difficult to model accurately in this shallow basin. Presence of large amounts of fine sediments and mud aggregates in the bay were not considered in these simulations, and these also affect the accuracy of calculated wave parameters. Tables 13a and 13b show the data range and errors in the calculated and measured wave height, period and direction. For this case with high wind speeds and shallow water depths, CMS-Wave was able to model the wind-waves with RMSE of 0.1 m in height (~25%) and 0.5 sec in period (~25%). Errors in wave direction were large, and are likely related to the difficulty in accurately measuring wave direction in low wave environments.

In general, wave calculations in a shallow basin are controlled by wind speed and energy loss due to white capping and a sensitivity analysis is warranted. This test case demonstrated that it is more efficient (twice faster) to run the CMS-Wave in a half-plane mode in a bay or lake alone application. In the case of a bay or estuary interacting with a sea through inlets/exits, it would be necessary to run the CMS-Wave in a full-plane mode.

4.2.2   Test C3-Ex2: Grays Harbor, Washington

Description: The purpose of this test case was to evaluate the combined wind and wave modeling capabilities of CMS-Wave in a large tidally-dominated inlet environment with an energetic wave climate. Extensive field data were collected in 1999, 2003 and 2005, and include wave and current measurements for Half Moon Bay, a region in the lee of the south jetty, in the navigation channel, north side of the channel, and back in the estuary, which provide good data to test a wave model (see Figures 28-32).

Grays Harbor (GH), located on the coast of southwest Washington, is one of the largest estuaries in the continental United States. The spring tidal prism reaches 570 million m^3 corresponding to the surface area of 200 km^2 at mean tide level, with a tidal range of 2.8 m. The entrance is approximately 2 km wide, and a deep-draft navigation channel is maintained at 12-13 m relative to mean low lower water. The entrance is protected by two rubble-mound jetties. The entrance to GH experiences extreme Northwest Pacific waves during winter. Significant wave heights commonly exceed 6 m during winter storms. Strong ebb currents that exist between the jetties can increase wave height by as much as 0.5 to 1.5 m as observed in the inlet entrance.

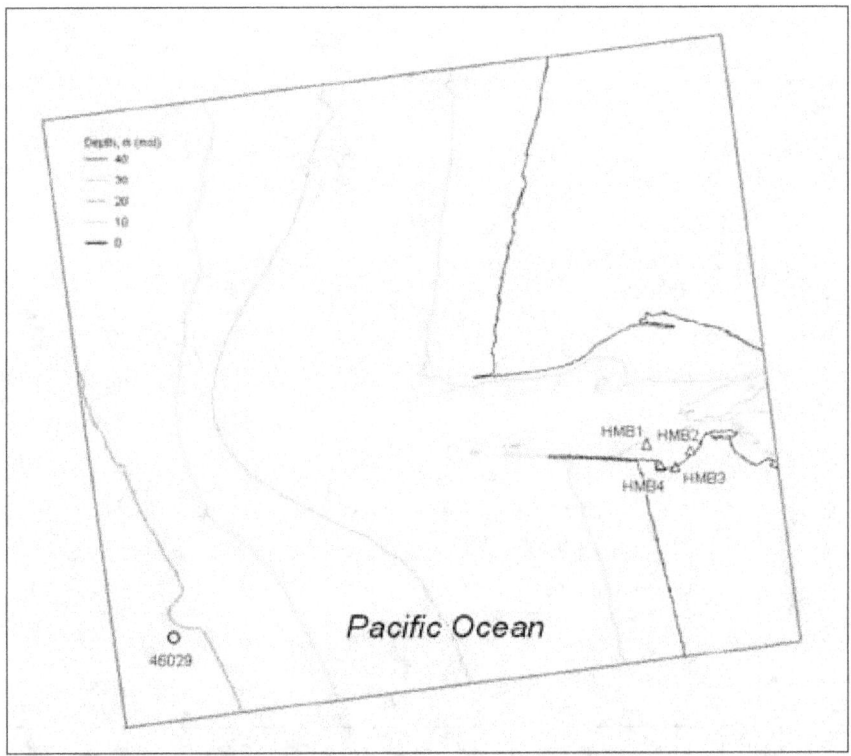

Figure 28. Wave data collection stations at Grays Harbor.

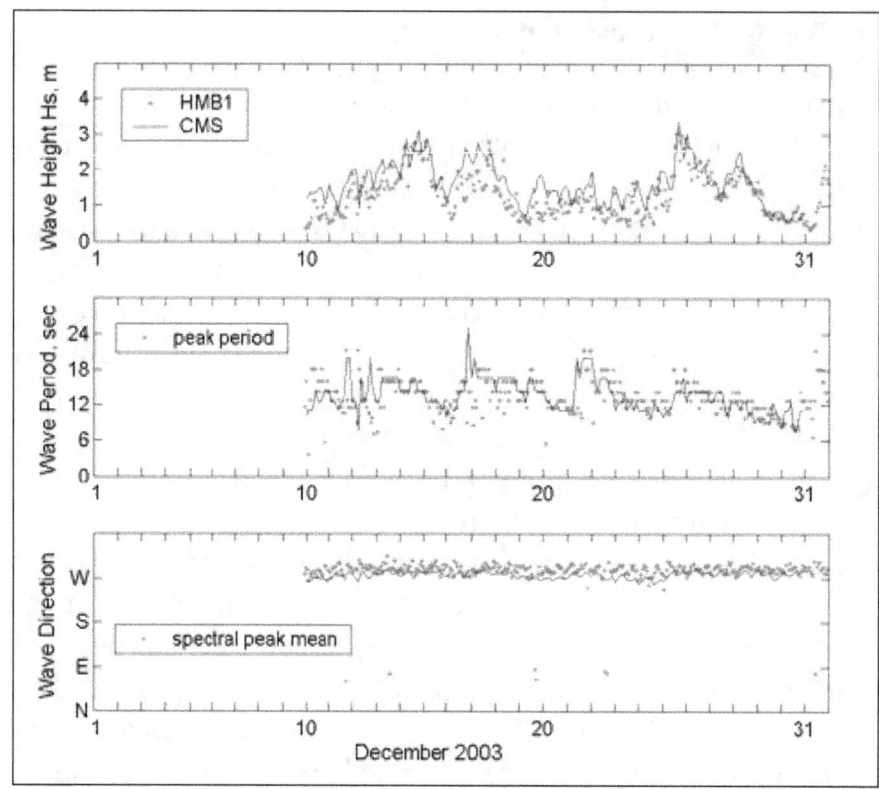

Figure 29. Measured and calculated waves at HMB1, 10-31 December 2003.

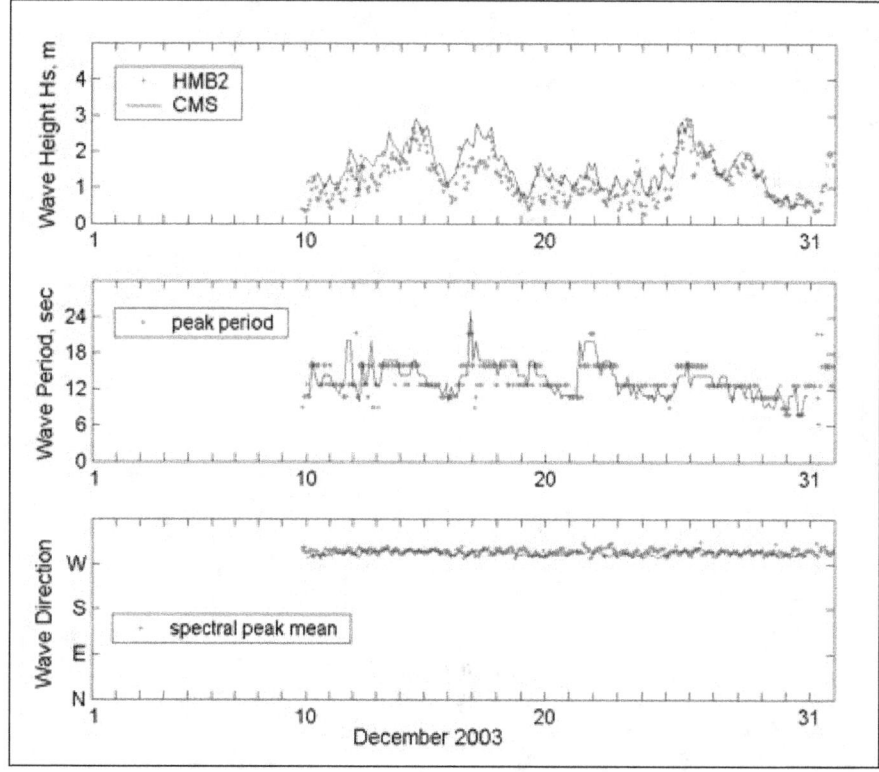

Figure 30. Measured and calculated waves at HMB2, 10-31 December 2003.

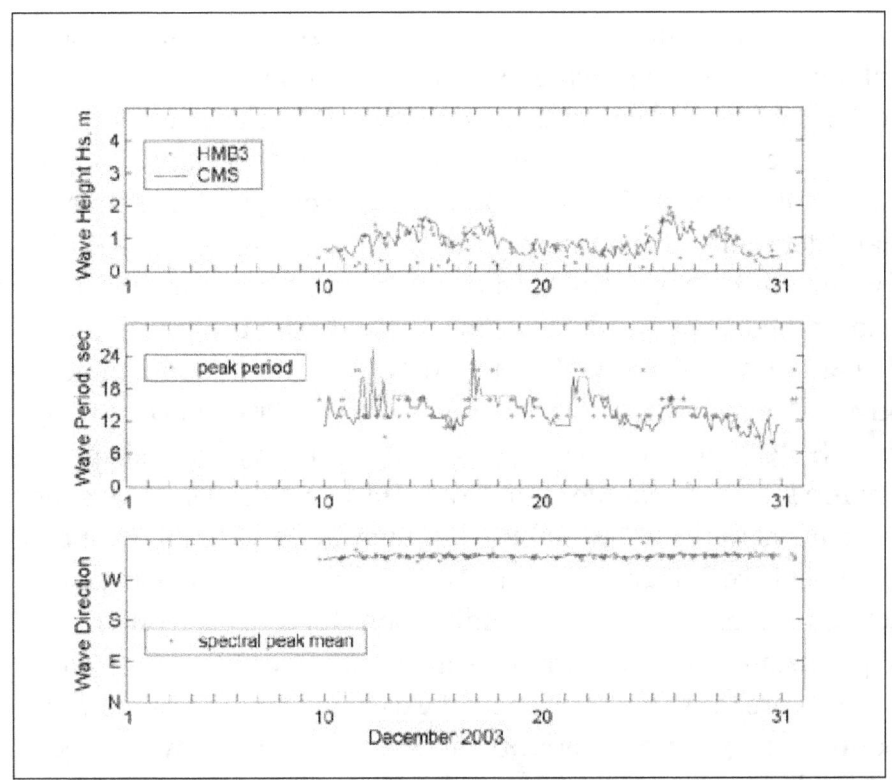

Figure 31. Measured and calculated waves at HMB3, 10-31 December 2003.

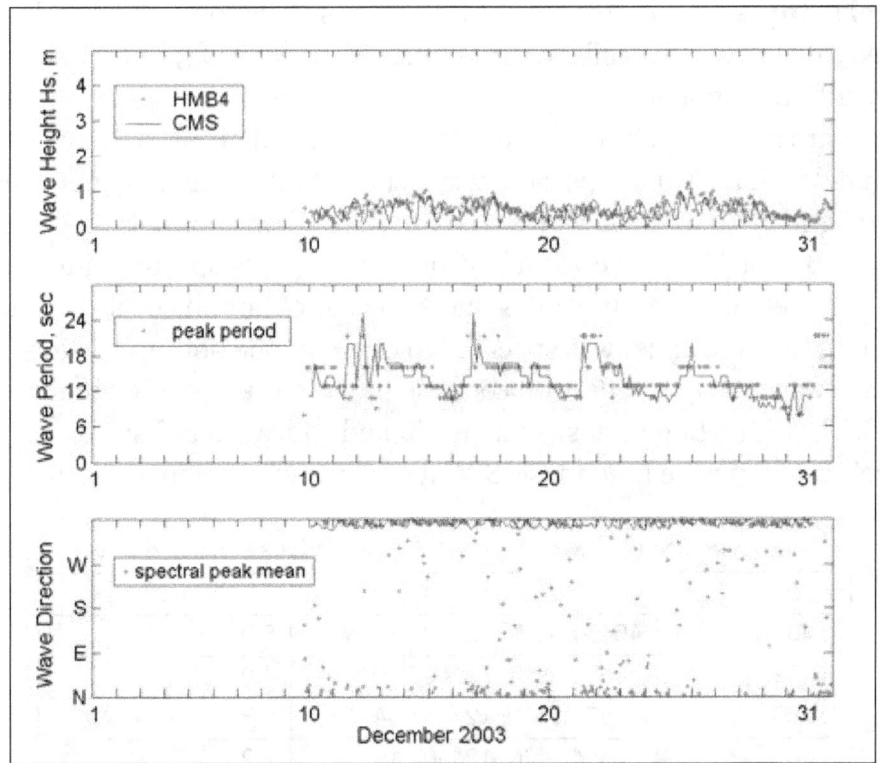

Figure 32. Measured and calculated waves at HMB4, 10-31 December 2003.

Strong wave refraction and diffraction at the eastern end of the south jetty contribute to increased beach erosion in Half Moon Bay. To examine the influence of waves and currents in Half Moon Bay, wave and current data were collected at four stations between December 2003 and February 2004, and Osborne and Davies (2004) provide details of instrumentation used and data collection and analyses. Table 14a provides the location of the four measurement stations. During the same time intervals, offshore wave information was available from a Coastal Data Information Program (CDIP) Buoy 036 (46°51.39'N, 124°14.67'W) in a water depth of 40 m (relative to Mean Tide Level, MTL) and from the National Data Buoy Center (NDBC) Buoy 46029 (46°8.63'N, 124°30.7'W), located approximately 100 km south-southeast of Grays Harbor. Ocean surface wind measurements were also available from Buoy 46029 (50 km west of Mouth of Columbia River). Figure 28 shows the location map and local data-collection stations (Buoy036 and Half Moon Bay stations HMB1 to HMB4). A winter storm occurred during 24-28 December 2003 with the largest offshore measured wave height exceeding 6 m. The simulations for the period of 10-30 December 2003 and this storm event were used in model validation.

Model setup and parameters: CMS-Wave simulations were conducted with a grid of 236 × 398 cells with variable cell spacing of 30 to 200 m (see Figure 28). Directional wave spectra from CDIP 036 served as the input at the seaward boundary. Wave generation, growth and propagation were computed with the wind input and wave-current interaction included on a spectral grid of 30 frequency bins (0.04 to 0.33 Hz with 0.01-Hz increment) and 35 direction bins (covering a half-plane with 5-deg spacing). Measured wind data were input to the model and a constant Manning coefficient of $n = 0.025$ (default value) was specified to calculate the bottom friction. A constant wave forward reflection coefficient of 0.5 was specified in the model. The diffraction intensity value of 4 and a forward reflection coefficient of 0.3 were used in CMS-Wave, which was run in a steering

Table 14a. Coordinates of wave monitoring stations at Grays Harbor.

Station	Coordinates	Depth, MTL (m)
HMB1	46o54'36"N, 124o07'30"W	8.0
HMB2	46o54'29"N, 124o06'50"W	4.0
HMB3	46o54'15"N, 124o07'04"W	1.8
HMB4	46o54'15"N, 124o07'18"W	2.2

Table 14b. Statistics of measured and calculated waves at Half Moon Bay, WA.

Station	Wave Parameter	RMSE	R	MAE
HMB1	Significant height (m)	0.51	0.80	0.41
	Peak period (sec)	3.3	0.34	2.1
	Mean direction (deg)	15.0	0.32	12.8
HMB2	Significant height (m)	0.48	0.80	0.38
	Peak period (sec)	2.4	0.59	1.7
	Mean direction (deg)	7.1	0.11	5.3
HMB3	Significant height (m)	0.35	0.55	0.27
	Peak period (sec)	5.1	0.37	2.9
	Mean direction (deg)	5.5	0.01	3.7
HMB4	Significant height (m)	0.27	0.24	0.22
	Peak period (sec)	4.8	0.45	2.7
	Mean direction (deg)	86	0.25	57

mode with the 3-hr interval with CMS-Flow to account for the influence of waves with current and tides. Other parameters not mentioned were set to the default values.

Results and Discussion: CMS-Wave was run to examine three different conditions:

a. Wave transformation only,
b. Wave transformation including local wind input, and
c. Coupling with CMS-Flow (with wind, tide, and wave-current interaction).

Results obtained with waves only and with wave plus wind forcings are similar, implying that waves offshore may have reached or were near the fully developed stage and mild to moderate wind speeds did not further promote wave growth.

The CMS simulations for GH were conducted from 10 to 30 December 2003. CMS-Wave was coupled with CMS-Flow on the same grid. The water level and flow boundary conditions of CMS-Flow were extracted from regional ADCIRC modeling, previously conducted by Demirbilek et al. (2010b). Modeling of infra-gravity waves, nonlinear wave-wave interactions, and wave transmission and overtopping of jetties was triggered in the wave simulation. Figures 29 to 32 show the measured and calculated significant wave heights, spectral peak periods, and mean directions at the four local data collection stations, HMB1 to HMB4, respectively. Table 14b provides the statistics between the measured and calculated wave heights, periods, and directions. The correlation between the measured and calculated significant heights is higher at HMB1 and HMB2 than HMB3 and HMB4. The correlation of calculated wave periods and directions to measurements is generally low, with the correlation coefficient less than 0.6, at all four HMB1 to HMB4. The RMSE and MAE for calculated wave heights are small, but period and direction errors are large and correlations are low, especially at HMB3 and HMB4 where wave diffraction, refraction, reflection, and breaking are significant. The scattering of wave direction at HMB1 and HMB4 as a result of wave reflection off the South Jetty may have caused the low correlation and high RMSE of model wave directions.

The trends in CMS-Wave calculation follow the data, but there are differences in the wave height, wave period and direction. The calculated wave results are more satisfactory with the coupled simulations of CMS-Wave and CMS-Flow as the measurements are the combination of waves and current.

The effect of shallower water on waves is evident in the comparison of calculated results with data at stations HMB3 and HMB4. The effect of current is more evident at stations HMB1 and HMB2, which are located in relatively deep water closer to the navigation channel. Results indicate CMS-Wave calculates wave height more accurately at HMB1 and HMB2 closer to the navigation channel in relatively deep water (8 and 4 m, respectively). The model results are less satisfactory at HMB3 and HMB4, situated in a sheltered area where wave diffraction, reflection, refraction, and wave shoaling, refraction and breaking in water the shallow (~2 m depth) are stronger as compared to HMB1 and HMB2.The results suggest that these mechanisms are not optimally modeled.

Overall, CMS-Wave performed reasonably well in this extremely dynamic and challenging field site. Because the modeling estimates would depend on wind, wave, tide and bathymetric inputs, a sensitivity study should be conducted to determine the best input parameters for future applications. For applications to jettied inlets with longer incident waves, the role of infra-gravity waves, nonlinear wave-wave interactions, and wave transmission and overtopping of breakwaters should be considered in CMS-Wave simulations to obtain reliable estimates.

4.2.3   Test C3-Ex3: Mouth of Columbia River, WA/OR

Description: The purpose of this test case was to validate CMS-Wave with data from the Mouth of Columbia River (MCR) entrance that is located at the WA/OR border. The MCR entrance area poses severe challenges to navigation because of its harsh climate (i.e., influence of winds, waves and tides). Severe storms and strong winds can occur unexpectedly, large waves impact the entrance in the fall and winter months, and the tidal range is high (2.1 m). These conditions cause sedimentation in the channel and along beaches, and damage to jetties protecting the shipping channel. The MCR entrance is one of the most dynamic sites in the Northwest region of the USA.

Directional wave measurements were collected by the U.S. Army Engineer District, Portland (Moritz 2005) between the north and south jetties from 1 August to 9 September 2005, at five monitoring stations. Table 15a gives coordinates and nominal depth, relative to mean tide level (MTL), of these monitoring stations. The incident wave spectrum was based on data from an offshore Buoy 46029 (46°7'N, 124°30.6'W) maintained by the National Data Buoy Center (NDBC) since 1984 (http://www.ndbc.noaa.gov). Figure 33 shows the model area bathymetric domain and wave gauge stations. Figure 34 shows sample time-series of wind and wave data collected from Buoy 46029 and Stations 4 and 5 at the MCR. The effects of waves interacting with tidal current at Stations 4 and 5 are clearly seen in the data as indicated by strong daily fluctuations of wave height, period and direction.

Model setup and parameters: The numerical grid covers a rectangular domain 20 km long (northing) and 35 km wide (easting), extending from the 128-m depth contour near Buoy 46029 to the entrance area of MCR (see Figure 33) with a constant resolution of 50 m. Wind forcing was taken from the buoy measurements adjusted to a 10-m elevation based on the 1/7 power law (Demirbilek et al. 2008). Wave generation and propagation were

Table 15a. Coordinates of wave monitoring stations at MCR (August-September 2005).

Station	Coordinates	Depth, MTL (m)
1	46°16'16"N, 124°03'23"W	9.7
2	46°15'47"N, 124°03'29"W	12.9
3	46°15'27"N, 124°03'13"W	21.7
4	46°15'04"N, 124°03'46"W	14.2
5	46°14'24"N, 124°03'58"W	10.4

Table 15b. Statistics of measured and calculated waves at the MCR.

Station	Wave Parameter	RMSE	R	MAE
Sta1	Significant height (m)	0.22	0.85	0.17
	Peak period (sec)	3.1	0.50	1.9
	Mean direction (deg)	12.4	-0.34	10.2
Sta2	Significant height (m)	0.19	0.70	0.15
	Peak period (sec)	3.0	0.47	1.8
	Mean direction (deg)	17.1	-0.43	13.7
Sta3	Significant height (m)	0.30	0.56	0.23
	Peak period (sec)	3.6	0.37	2.1
	Mean direction (deg)	32.3	-0.06	25.6
Sta4	Significant height (m)	0.29	0.77	0.20
	Peak period (sec)	2.3	0.56	1.4
	Mean direction (deg)	20.8	0.30	15.0
Sta5	Significant height (m)	0.37	0.75	0.24
	Peak period (sec)	1.8	0.48	1.1
	Mean direction (deg)	20.2	0.22	14.9

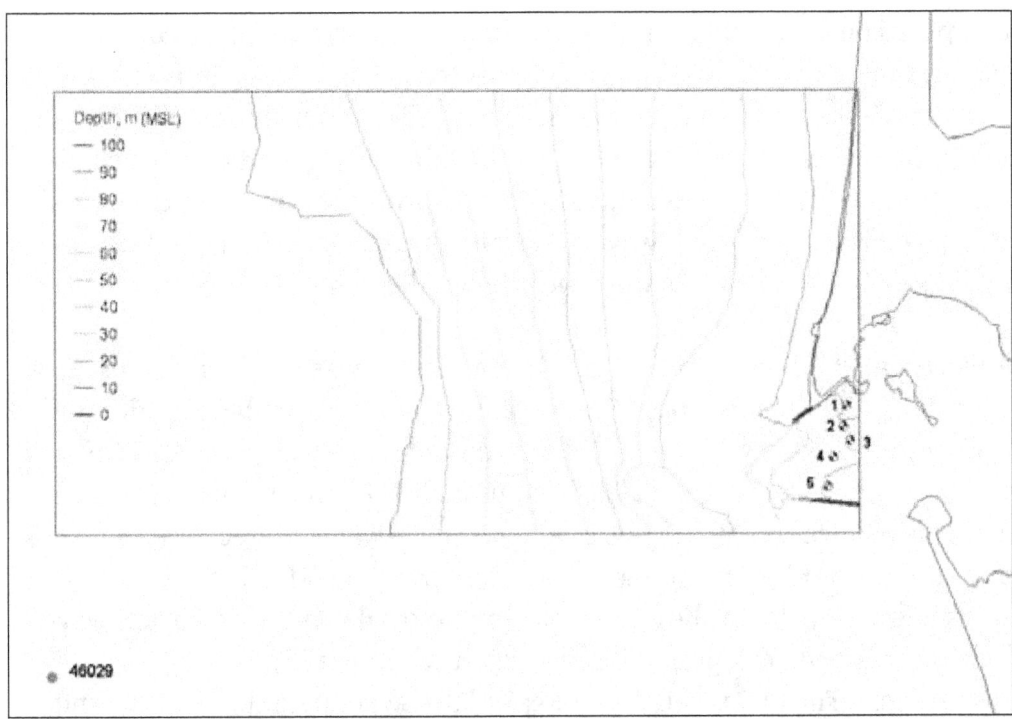

Figure 33. Wave model domain and directional wave data collection locations.

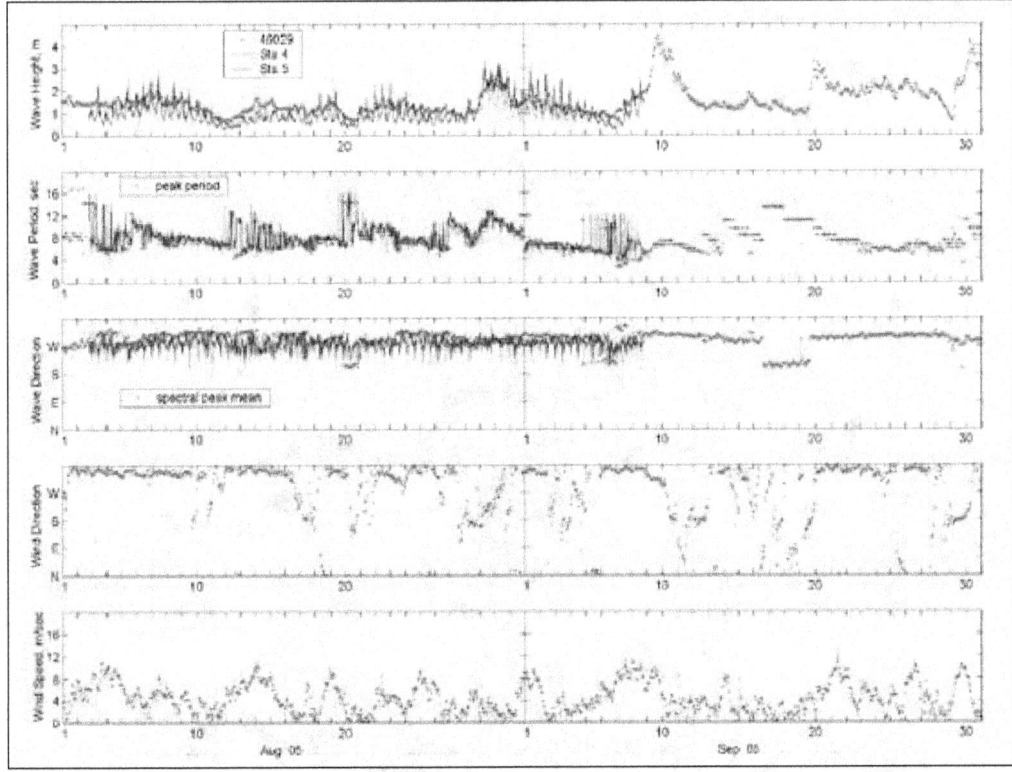

Figure 34. Wave and wind data collected at Buoy 46029, Stations 4 and 5.

computed on a spectral grid of 30 frequency bins (0.04 to 0.33 Hz with 0.01-Hz increment) and 35 direction bins (covering a half-plane with 5-deg spacing). Measured wind data were input to the model and a constant Manning coefficient of $n = 0.025$ (default value) was specified to calculate the bottom friction. A constant value of forward reflection coefficient (0.5) was specified. Wind input, wave diffraction, wave-current interaction, and infra-gravity wave features were included in these simulations.

Results and Discussion: The CMS simulation was conducted for 1 August to 9 September 2005. CMS-Wave was coupled with CMS-Flow on the same grid. The water level and flow boundary conditions of CMS-Flow were extracted from regional ADCIRC modeling (Demirbilek et al. 2008). The role of infra-gravity waves, nonlinear wave-wave interactions, and wave transmission and overtopping breakwaters were included in these simulations. Figures 35 to 39 show the measured and calculated significant wave heights, spectral peak periods, and mean directions at Sta 1 to 5, respectively. The error statistics for calculated wave heights, periods, and directions are listed in Table 15b.

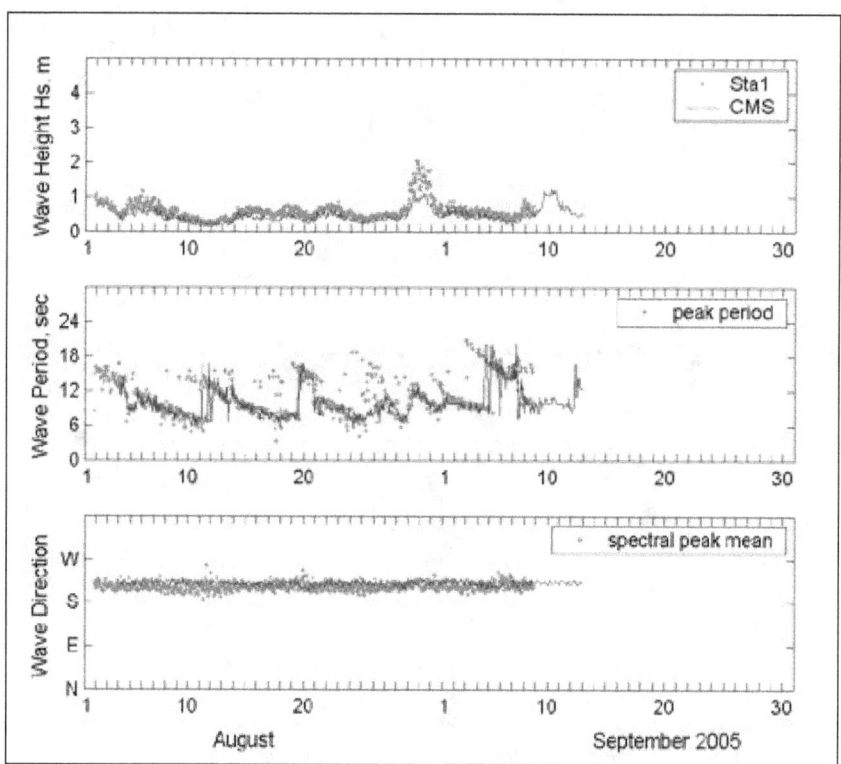

Figure 35. Measured and calculated waves at Sta1, 1 August - 9 September 2005.

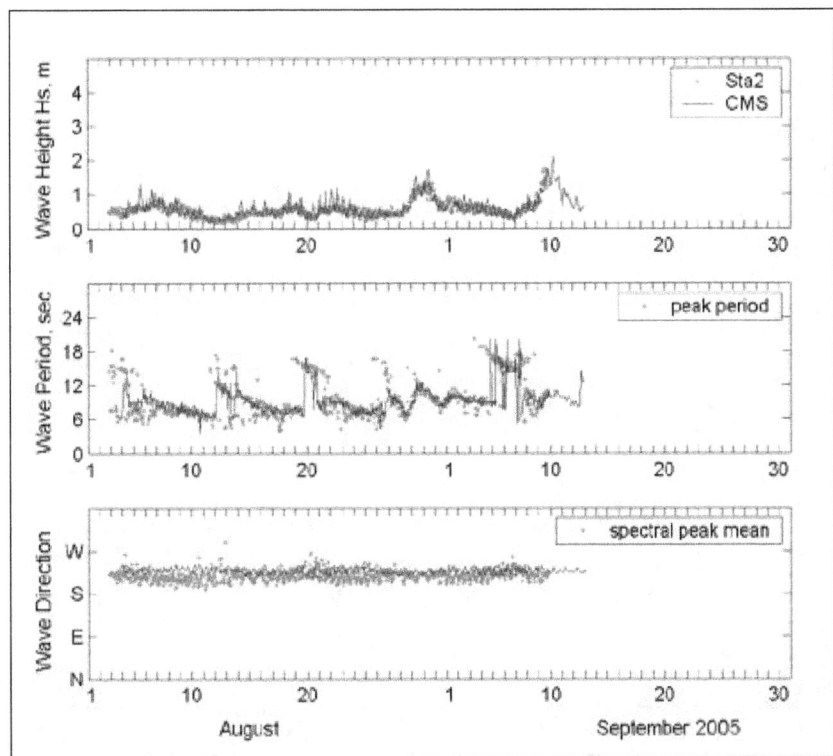

Figure 36. Measured and calculated waves at Sta2, 1 August - 9 September 2005.

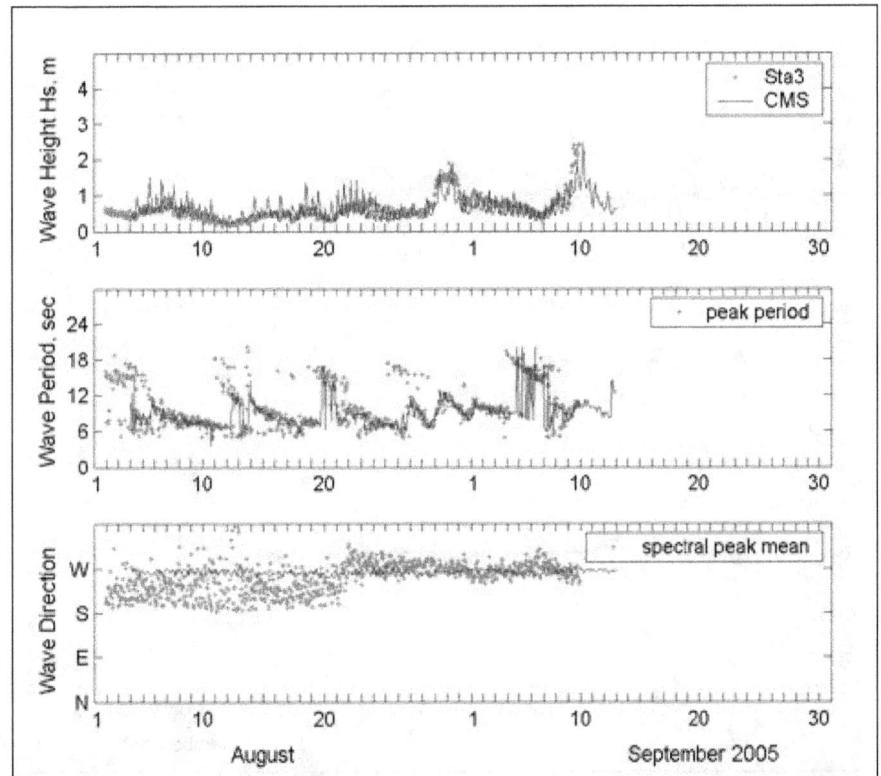

Figure 37. Measured and calculated waves at Sta3, 1 August - 9 September 2005.

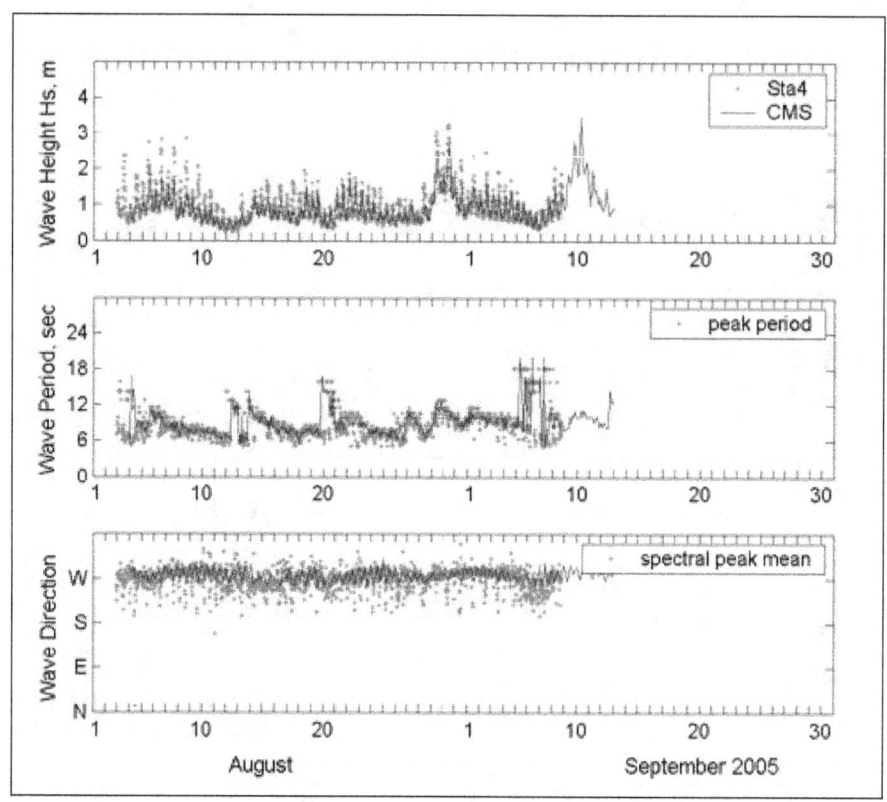

Figure 38. Measured and calculated waves at Sta4, 1 August - 9 September 2005.

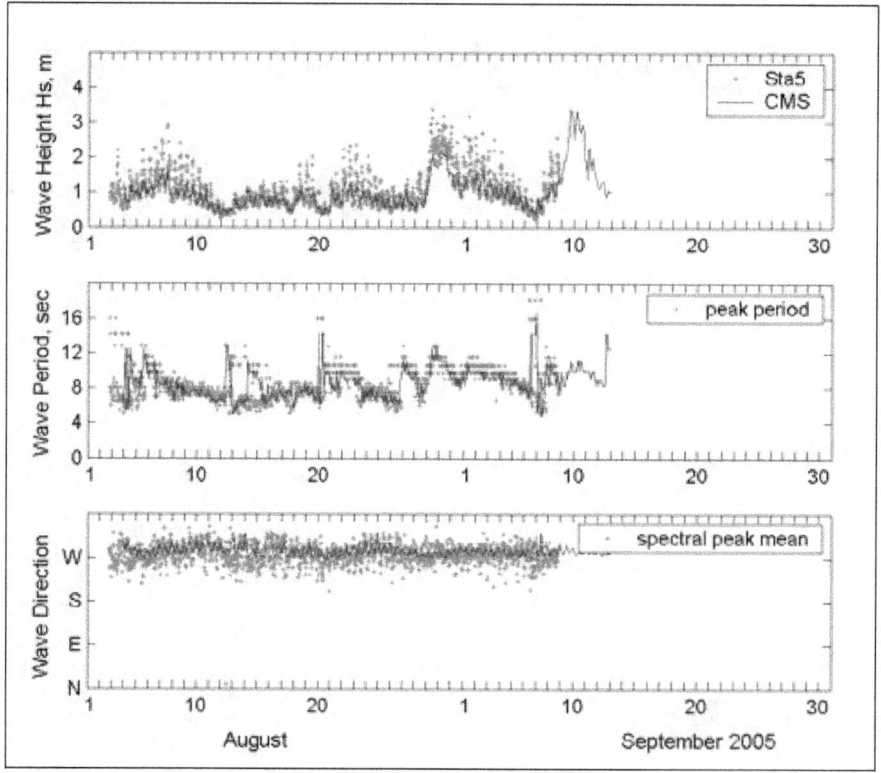

Figure 39. Measured and calculated waves at Sta5, 1 August - 9 September 2005.

Strong tidal and river currents are observed at the MCR, with a typical average peak current magnitude of 2 m/sec. Large wave events occurred at 1000 GMT on 7 August, at 0000 GMT on 30 August, and at 1800 GMT on 9 September 2005 (see Figure 34). The effects of waves interacting with tidal current at these stations are depicted in the data by strong daily fluctuations of wave height, period and direction.

Comparison at the offshore buoys and nearshore gauges indicate that waves experience significant changes in their transformation from deep to shallow water. The CMS-Wave model predictions were validated with data obtained from field experiments using measured wave heights, periods, and directions. Because two different types of Acoustic Doppler Current Profiler (ADCP) were used in the local data collection, the wave and current data from Sta 4 and 5 are more reliable than from Sta 1 to 3. Statistics at Sta 4 and 5 show overall a better correlation with data and comparatively smaller RMSE and MAE errors between measured and calculated wave heights, periods and directions. CMS-Wave calculations of wave height are in better agreement for Sta 4 and Sta 5, located closer to the navigation channel in relatively deep water. Maximum error in wave heights was less, about 20 percent, but period and direction correlations were weak at all measurement stations. Errors for wave heights averaged 0.6 m, for wave direction 15 deg, and for period 0.7 sec.

As in the previous example for a West coast site, CMS-Wave applications to jettied inlets with incident longer period waves would benefit from including processes such as infra-gravity waves, nonlinear wave-wave interactions, and wave transmission and overtopping of breakwaters.

4.2.4   Test C3-Ex4: Southeast Oahu Coast, Hawaii

Description: The purpose of this test case was to check the capability of CMS-Wave for producing reliable wave predictions on fringing reefs in the nearshore using field data from Oahu Island, Hawaii. Directional wave data were collected at the southeast coast of Oahu for the Southeast Oahu Regional Sediment Management demonstration project conducted by the U.S. Army Engineer District, Honolulu. The data collection equipment included three Acoustic Doppler Velocimeters (ADV) installed from 9 August to 14 September 2005, in the nearshore (Cialone et al. 2008). Table 16a lists the location of the ADVs. The corresponding offshore wave data are available from a CDIP Buoy 098 (21°24.9'N, 157°40.7'W), deployed near the study site. This data collection period was dominated by

Table 16a. Coordinates of ADV stations at southeast Oahu.

Station	Coordinates	Depth, MTL (m)
ADV1	21°23'52"N, 157°43'05"W	2.5
ADV2	21°22'31"N, 157°42'14"W	2.7
ADV3	21°19'48"N, 157°40'56"W	2.5

Table 16b. Statistics of measured and calculated waves at Southeast Oahu.

Station	Wave Parameter	RMSE	R	MAE
ADV1	Significant height	0.62 (m)	0.82	0.33 (m)
	Peak period	3.1 (sec)	0.41	2.5 (sec)
	Mean direction (deg)	13.2 (deg)	0.35	10.1 (deg)
ADV2	Significant height	0.41 (m)	0.83	0.44 (m)
	Peak period	2.5 (sec)	0.55	2.0 (sec)
	Mean direction	7.5 (deg)	0.22	5.5 (deg)
ADV3	Significant height	0.45 (m)	0.51	0.33 (m)
	Peak period	4.3 (sec)	0.41	2,5 (sec)
	Mean direction (	6.5 (deg)	0.33	4.9 (deg)

trade winds, typically occurring from April through September in Hawaii, and was characterized by wind consistently blowing from the northeast. The ocean surface wind was measured at NDBC Buoy 51001 (23°25.92'N, 162°12.47'W), approximately 250 km northwest of Oahu Island. Water level data are available from two near NOAA stations: Station 1612340 (21°18.4'N, 157°52'W) at Honolulu Harbor, and Station 1612480 (21°26.2'N, 157°47.6'W) at Kaneohe Bay. Figure 40 shows the ADV, CDIP buoy, and NOAA tidal station locations.

Model setup and parameters: For CMS-Wave simulations, a nearshore bathymetry grid was developed covering a 24.2-km coastline including Mokapu Point, Kailua Bay and Waimanalo Bay. The seaward boundary extends to the 300-m contour, with a maximum 510-m depth (Figure 40). The grid consisted of 310 × 968 cells with cell size of 25 m × 25 m. The incident wave two-dimensional spectra at the offshore boundary were obtained from the CDIP Buoy 098. Both wind and water level data were input to the simulation. Thirty frequency bins (0.04 to 0.33 Hz with 0.01-Hz increment) and 35 direction bins (covering a half-plane with 5-deg spacing) were specified for the wave calculation. The default parameters

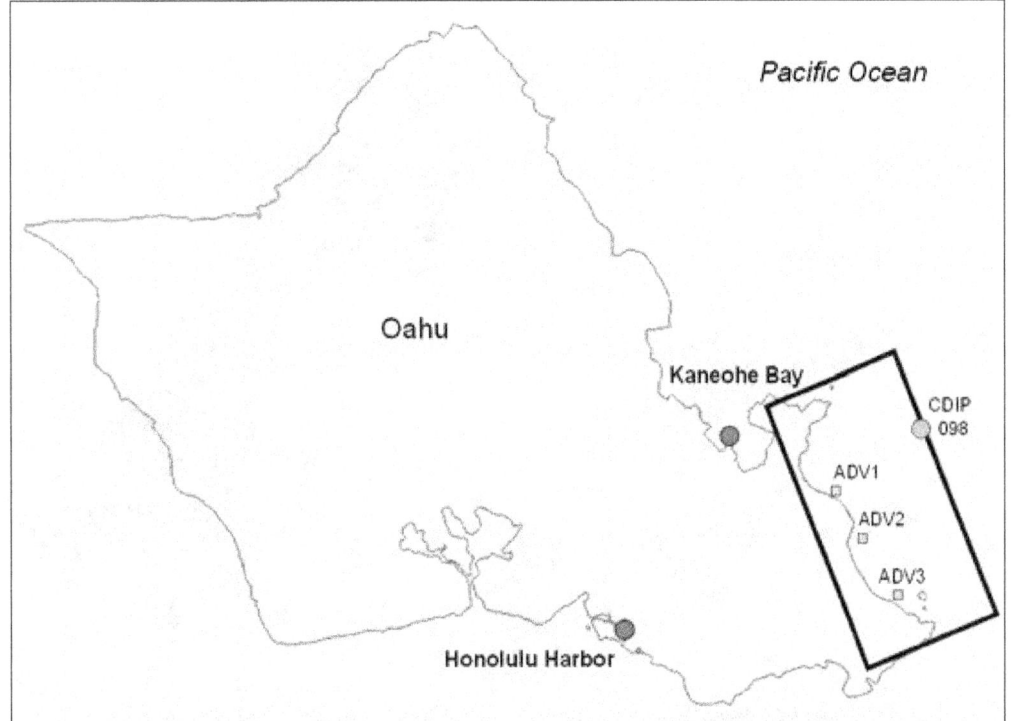

Figure 40. Map of CMS-Wave model domain with Tide and ADV stations.

were used in these simulations unless otherwise noted. For the southeast Oahu coast, it is necessary to apply large bottom friction coefficients to simulate effects of the offshore reefs. Because the Darcy-Weisbach friction coefficient (C_f) is physically valid for only a small range between 0 and 0.05, CMS-Wave will automatically use the Manning coefficient if the user specified value is greater than 0.05. Figure 41 shows different bottom friction coefficients (0.01 to 0.16) specified in the computational domain. The simulation was run with both forward and backward reflection. A constant reflection coefficient value of 1 and default diffraction intensity value of 4 were selected. Surface wind input and water level data were input in addition to the offshore spectral wave forcing.

Results and Discussion: Because the southeast Oahu coast is fronted by an extensive reef system, waves approaching the shore dissipate more energy than if traveling over a sandy bed. It was necessary to specify different bottom friction coefficients in the reef and non-reef areas to account for the reef roughness.

Figures 42 and 43 show calculated wave results with data measured at the three nearshore ADV locations. The calculated wave heights agree with the measurements, but there are also some significant differences. There is

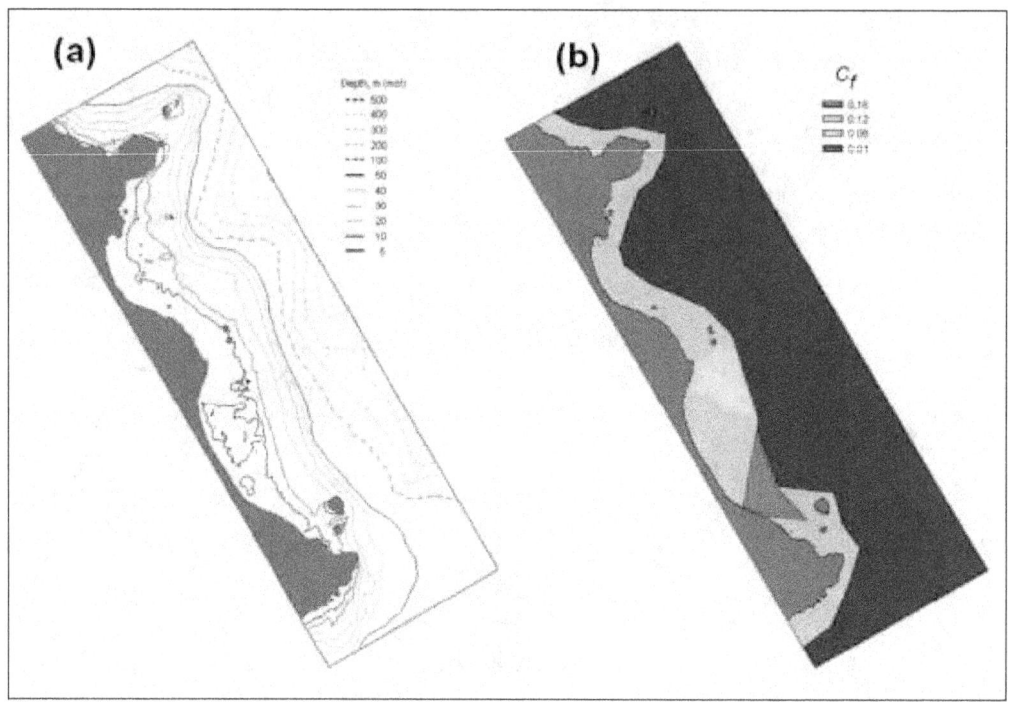

Figure 41. (a) Bathymetry grid, and (b) different bottom friction coefficient regions.

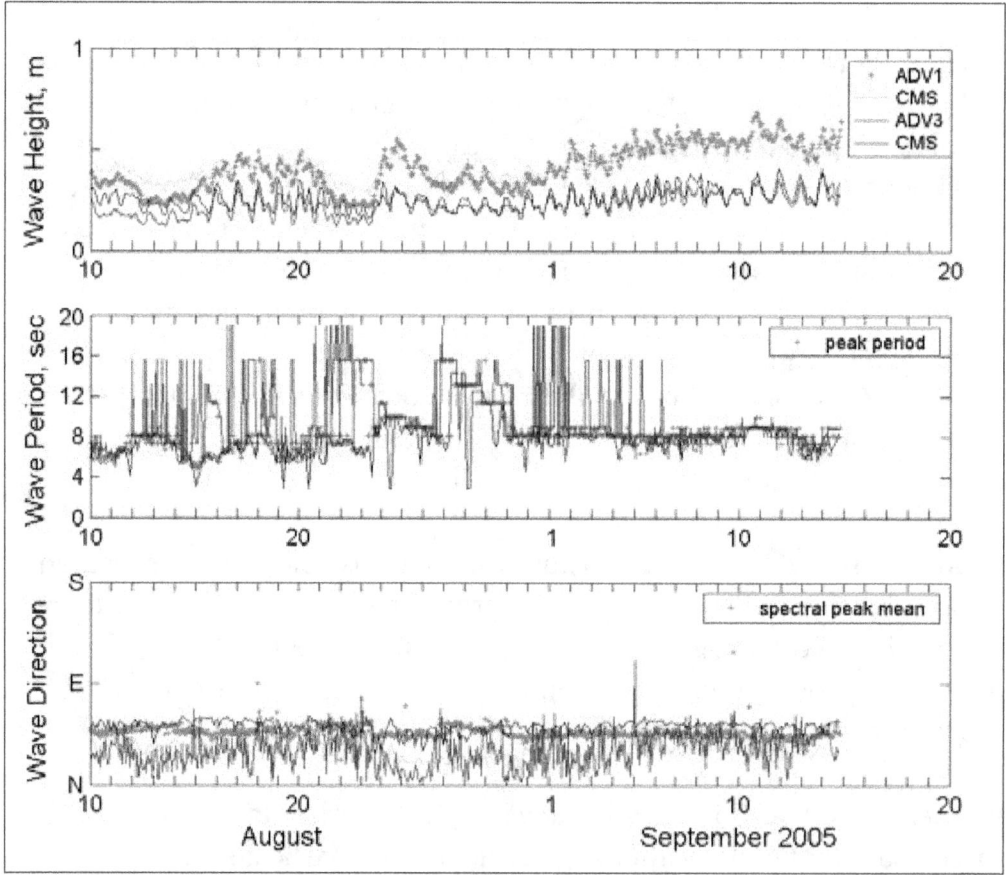

Figure 42. Measured and calculated waves at ADV1 and ADV3, August-September 2005.

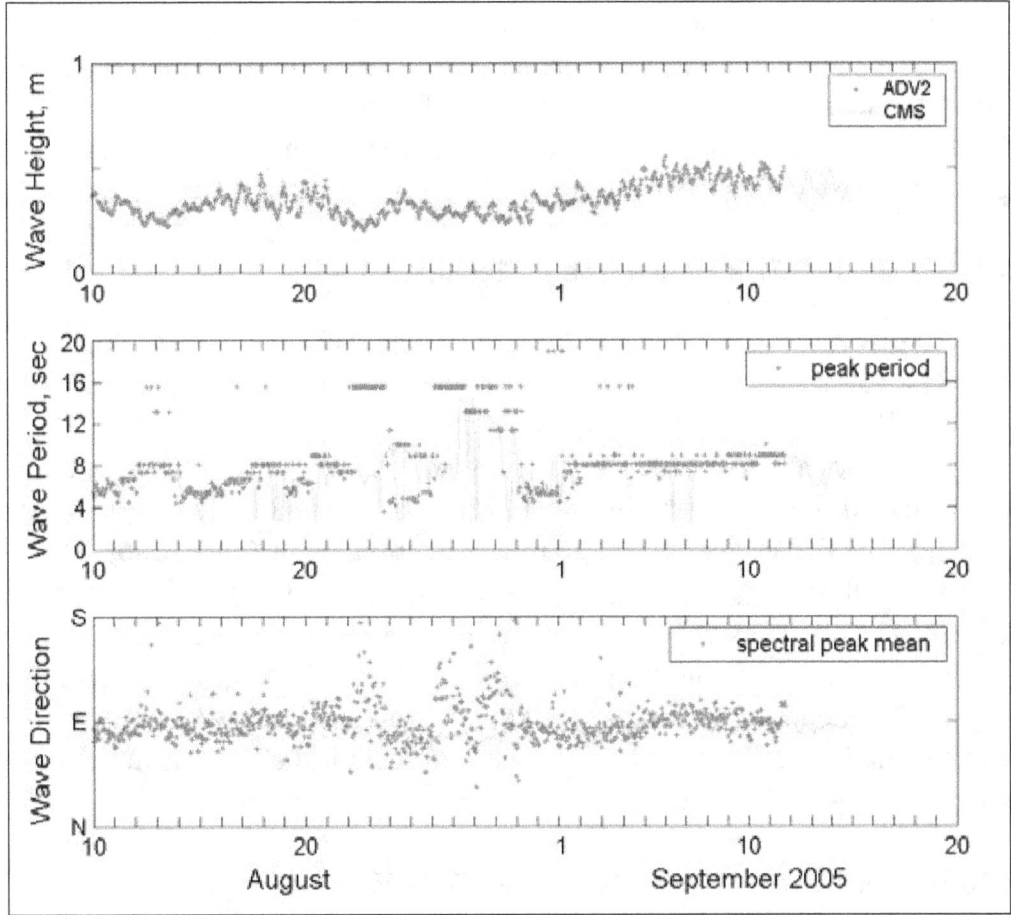

Figure 43. Measured and calculated waves at ADV2, August-September 2005.

considerable noise in wave period and direction measurements from ADV2 and ADV3 where the local bathymetry and reef bottom may influence the approaching waves. Table 16b presents the statistics for comparison between model results and data. Wave height correlations for Sta1 and Sta2 are good, but other statistics are not as favorable. For example, RMSE for wave period at Sta2 is 2.4 sec, and RMSE for the wave height at other stations is nearly on the same order as the wave height. These differences may be caused by gauge problems or local processes which were not properly resolved by the model. Additional data from reef environments are needed to further evaluate the model's skills for wave prediction over complex reef bathymetries with different wave conditions.

For this wave height calculation over a reef, CMS-Wave results agreed satisfactorily with the field measurements. In similar applications, successful model performance may require proper calibration of the model with the site specific data to determine an applicable bottom friction coefficient. Wave height, period and nonlinearities as waves pass over reefs, seabed roughness, surface irregularities, and reef face slopes can affect modeling results significantly. A careful sensitivity analysis should be conducted to assess the effects of these processes and the associated parameters on model predictions.

4.2.5   Test C3-Ex5: Field Research Facility, NC

Description: The USACE Field Research Facility (FRF) at Duck, North Carolina, has collected long-term wave data along a cross-shore wave array and two Waverider buoys. The array has four bottom mounted Nortek Acoustic Wave and Current (AWAC) sensors at depths of 5, 6, 8 and 11 m and two directional Waverider buoys at 17-m and 26-m depths (Figures 44 and 45). The Waverider buoy at 26 m was maintained by CDIP, Buoy 430, available online at http://cdip.ucsd.edu. The wind measurements are available from NOAA coastal Station 8651370 at the end of the FRF Pier and from a National Data Buoy Center (NDBC) directional wave Buoy 44014 at 48-m depth. The array and buoys spanned 95-km cross-shelf to capture the wave transformation processes from the outer continental shelf to within the surf zone (Hanson et al., 2009).

Model Setup and Parameters: The CMS-Wave grid covered a rectangular area extending 16 km from CDIP 430 to shoreline and 19 km alongshore. The domain included variable cell size in the cross-shore direction, ranging from 5 m in the nearshore to 250 m offshore and

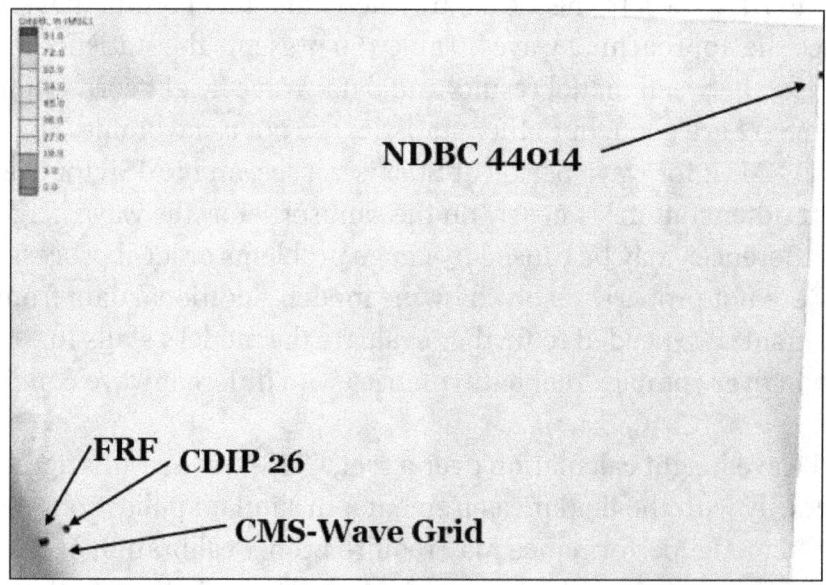

Figure 44. Offshore bathymetry showing NDBC and CDIP buoy locations.

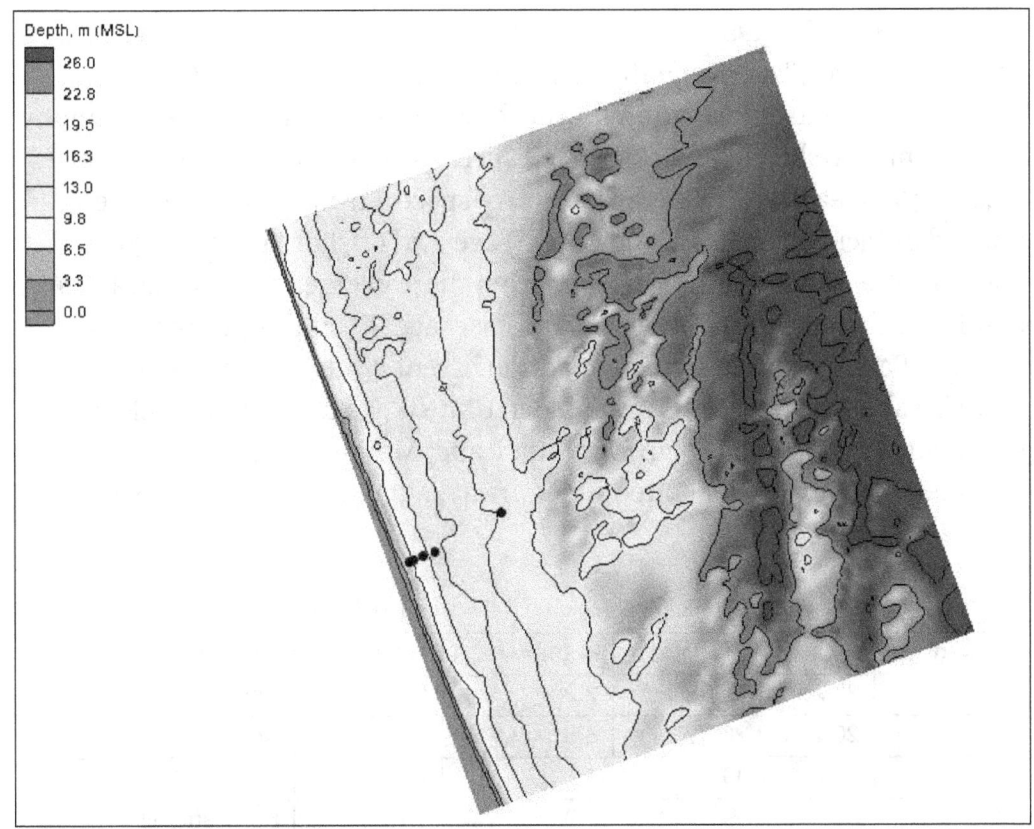

Figure 45. CMS-Wave modeling domain with bathymetric contours.

constant cell size of 70 m applied in the alongshore direction. Bathymetry data in the vicinity of the wave array from the shore to 15 m below MTL was obtained from an FRF survey in June 2010. Seaward of the 15 m contour, historic NOAA survey data were applied. Wetting and drying, nonlinear wave effects, and infragravity wave effects were active while forward and backward reflection were inactive in all model runs. The default parameters were used for other parameters.

Both wind measurements at the FRF Pier and at NOAA buoy 44014 were tested. Sensitivity to bottom friction was tested with a constant Darcy-Weisbach bottom friction coefficient $C_f = 0.01$ representing an upper bound to determine the model's sensitivity to bottom friction, relative to simulations without bottom friction.

Four storm events were run. Events 1 to 3 were run with the FRF Pier wind measurements only. Event 4 was also tested without wind forcing to investigate model sensitivity to different wind measurement location. All events (Events 1-4) included time varying water level measured at the FRF Pier.

Results and Discussion: The four storm events included three north-easters and 2009 Hurricane Bill, investigated by Hanson et al. *(*2009). Table 17a presents information for the four storm events with the maximum significant wave height and associated spectral peak period measured at CDIP 430 for each storm. Measured wave spectra from CDIP 430 were applied as incident waves at the CMS-Wave grid offshore boundary; the 180° spectra were directly excerpted from the 360° measured spectra using a utility included within the CMS-Wave package. Events 1-3 were run with only the Battjes and Janssen formulation. Event 4 was run with all four wave breaking formulations available in CMS-Wave to assess model sensitivity for large waves during a hurricane. Measured waves applied at the model boundary are plotted in Figure 47, along with winds measured at the FRF pier.

Table 17a. Events selected for wave model validation.

Event	Date	H_s (m)	T_p (sec)	Description
1	27-28 Sep 2008	2.2	12	Northeaster
2	20-21 Oct 2008	3.3	14	Northeaster
3	19 Feb 2009	2.1	14	Northeaster
4	22 Aug 2009	3.9	18	Hurricane Bill

Table 17b. Statistics for all stations and events.

Statistics	H_s (m)	T_p (sec)	Direction (deg)
RMSE	0.32	1.50	7.31
R	0.90	0.77	0.86
MAE	0.24	1.06	5.93

Table 17c. Wave height statistics for breaking formula sensitivity, Event 4.

Statistics	B&J	Goda	C&K	Miche
RMSE, m	0.39	0.39	0.40	0.39
R	0.86	0.86	0.87	0.86
MAE, m	0.32	0.32	0.32	0.32

Table 17d. Wave height statistics for wind sensitivity, Event 4.

Statistics	NOAA Buoy	FRF Pier	No Wind
RMSE, m	0.43	0.39	0.37
R	0.86	0.86	0.87
MAE, m	0.34	0.32	0.30

Table 17e. Sensitivity of wave height statistics to bottom friction.

Statistics	Event 1		Event 2		Event 3		Event 4	
	C_f =0.0	C_f =0.01	C_f =0.0	C_f =0.01	C_f =0.0	C_f =0.01	C_f =0.0	C_f =0.01
RMSE, m	0.17	0.28	0.32	0.27	0.25	0.18	0.39	0.61
R	0.10	0.07	0.95	0.94	0.55	0.49	0.86	0.84
MAE, m	0.14	0.25	0.25	0.21	0.17	0.16	0.32	0.50

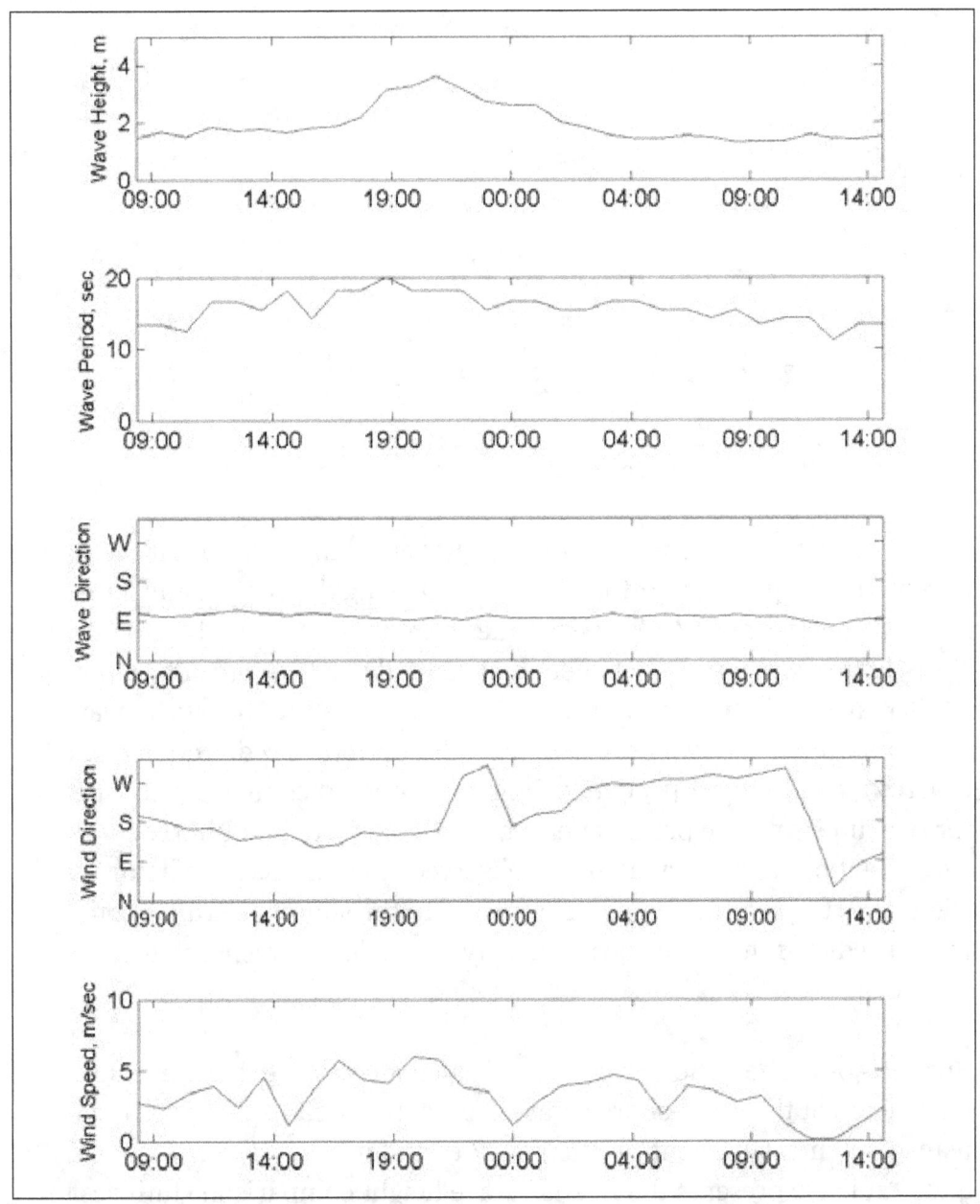

Figure 46. Input wave and wind conditions during Hurricane Bill.

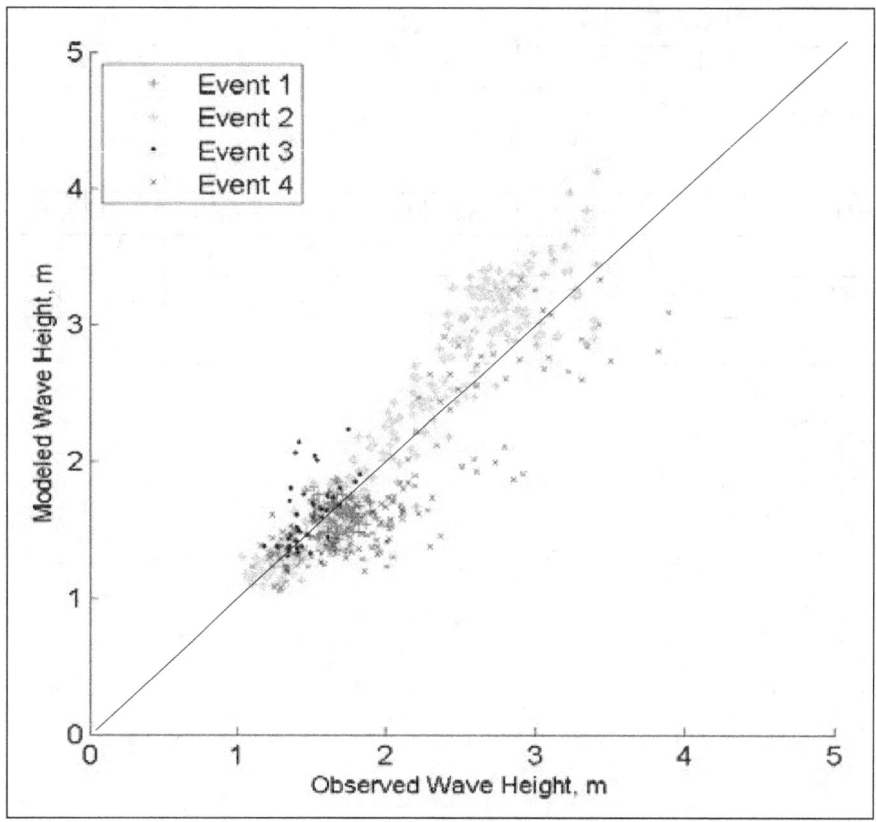

Figure 47. Significant wave height for all stations and events.

Figure 47 plots the calculated wave height versus the data for all events (Events 1-4) neglecting bottom friction and using the Battjes and Janssen formulation and the FRF Pier wind measurements. Table 17b presents the statistics for comparison between model and data. These results show that CMS-Wave predictions compare well to data at the FRF for storm waves, with correlation coefficients of wave height, period, and direction equal to 0.9, 0.77, and 0.86, respectively. Figures 48 and 49 compare wave spectra for two different time periods near the peak of Hurricane Bill. The agreement between calculated and observed spectra was satisfactory, with slightly better agreement occurring closer to the shore. Overall, the model underpredicted the wave energy density at the deeper measurement locations.

Figures 50-54 show the measured and calculated significant wave height for Event 4 at the cross-shore array 17-m to 5-m stations, respectively, using wind measurements at the FRF Pier and incident waves from CDIP 430. Table 17c presents the Event 4 wave height comparison statistics for different breaking formulas. These statistics indicate that the choice of breaker formula has a minimal effect on calculated wave height.

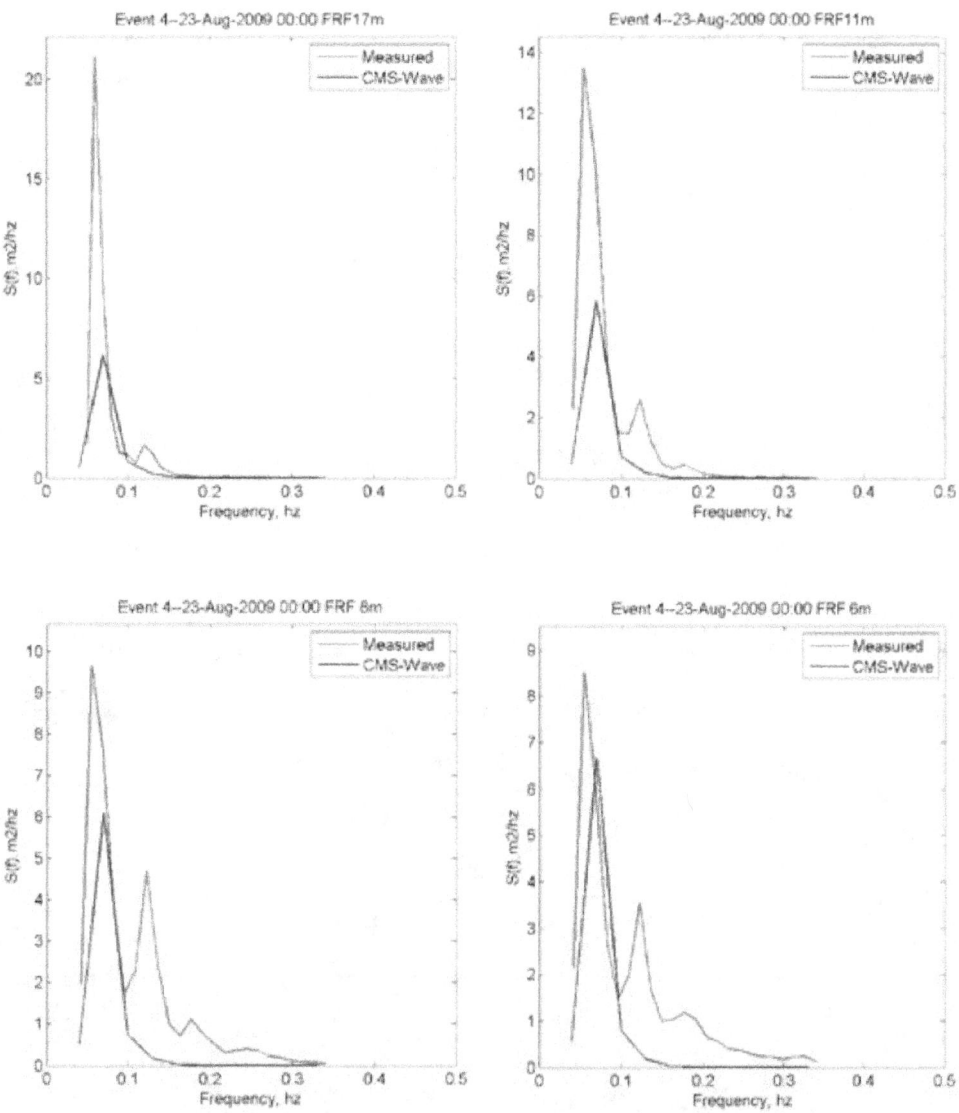

Figure 48.Measured and calculated wave spectra for 17, 11, 8, and 6 m stations at 0000 GMT, 23 August 2009 during Hurricane Bill (Event 4).

Figure 55 compares the calculated wave height at the 11-m station for Event 4 in the conditions without wind, with wind measurements at the FRF Pier, and with wind measurements at the NDBC buoy. Table 17d presents corresponding statistics with varying wind forcing. The no wind condition yielded the lowest RMSE and MAE but slightly higher correlation coefficients as compared to with wind condition. Table 17e presents the statistics for the sensitivity test of neglecting the bottom friction and with bottom friction using a constant $C_f = 0.01$ in all event (Events 1-4). Neglecting bottom friction tends to fit slightly better to the data.

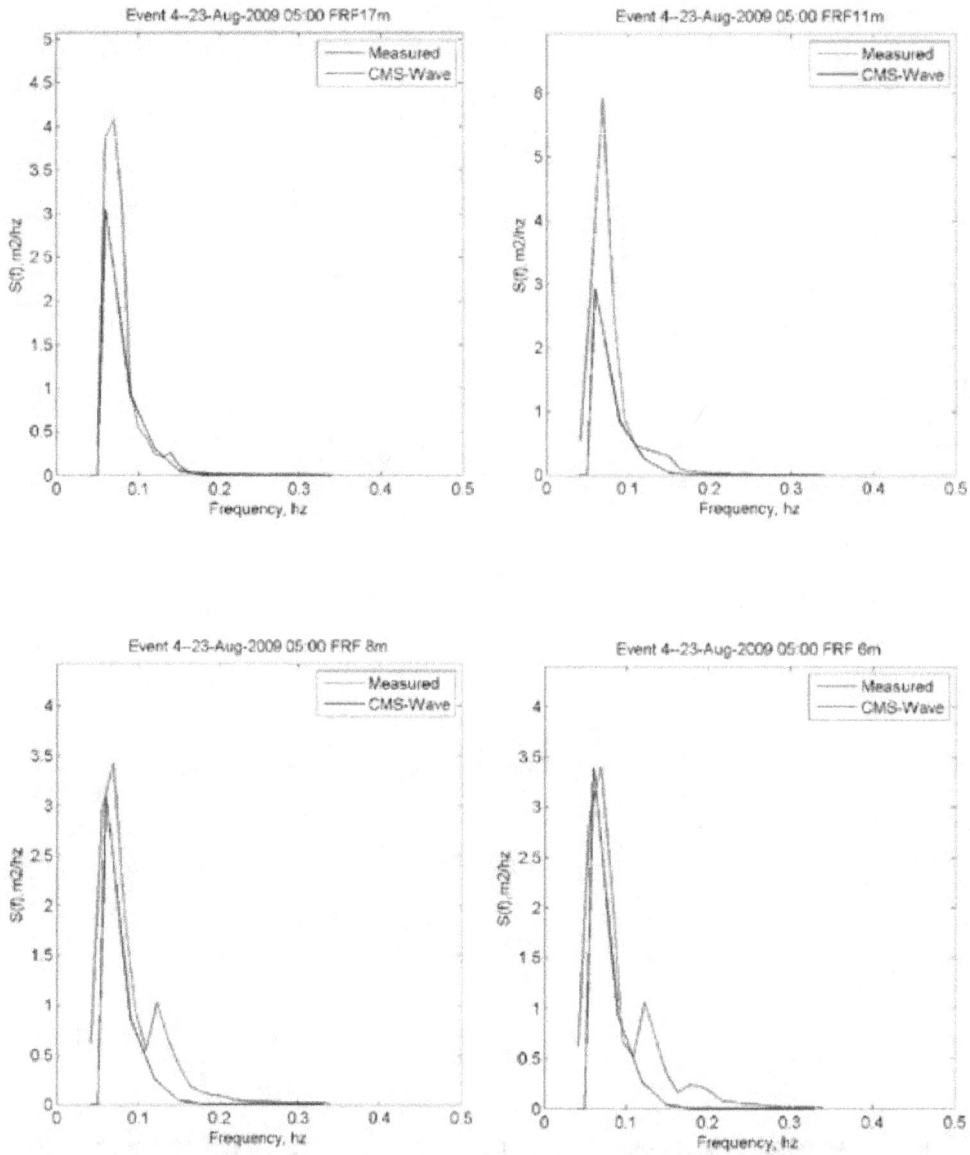

Figure 49.Measured and calculated wave spectra for 17, 11, 8, and 6 m stations at 0500 GMT, 23 August 2009 during Hurricane Bill (Event 4).

CMS-Wave was not sensitive to different available breaking formulas applied in the nearshore wave transformation at the FRF during the storm events. Tests indicated that CMS-Wave predictions were sensitive to the value of bottom friction used in simulations, resulting in a 25% or higher difference in RMSE and MAE (most readily observed during Hurricane Bill, a storm with large swell). The best model performance is obtained by neglecting bottom friction while applying the Battjes and Janssen breaking formula with input wind measured at the FRF Pier and incident waves from CDIP 430.

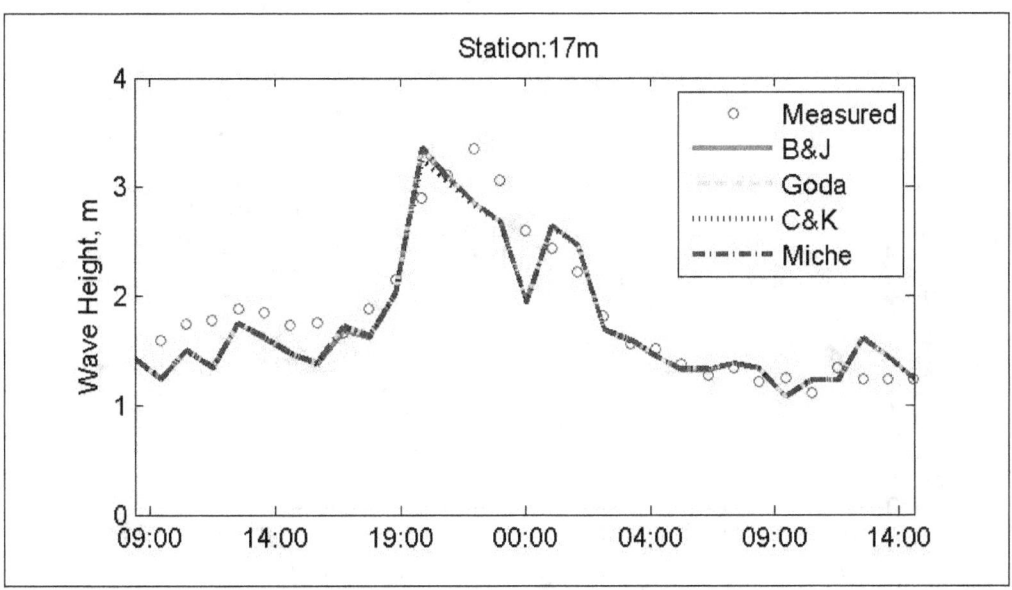

Figure 50. Calculated and measured wave height at 17-m station for Event 4, 22-23 August 2008.

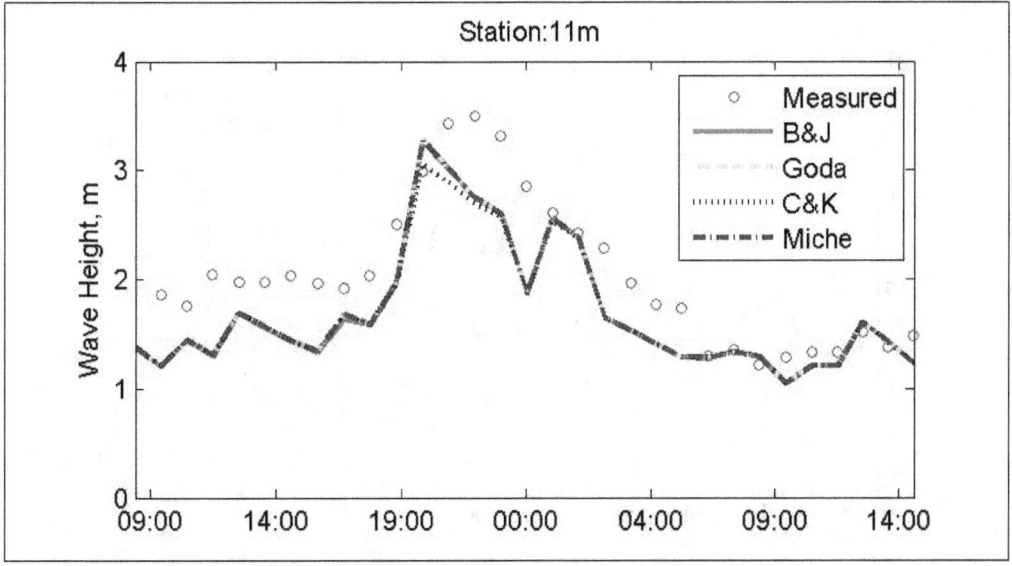

Figure 51. Calculated and measured wave height at 11-m station for Event 4, 22-23 August 2008.

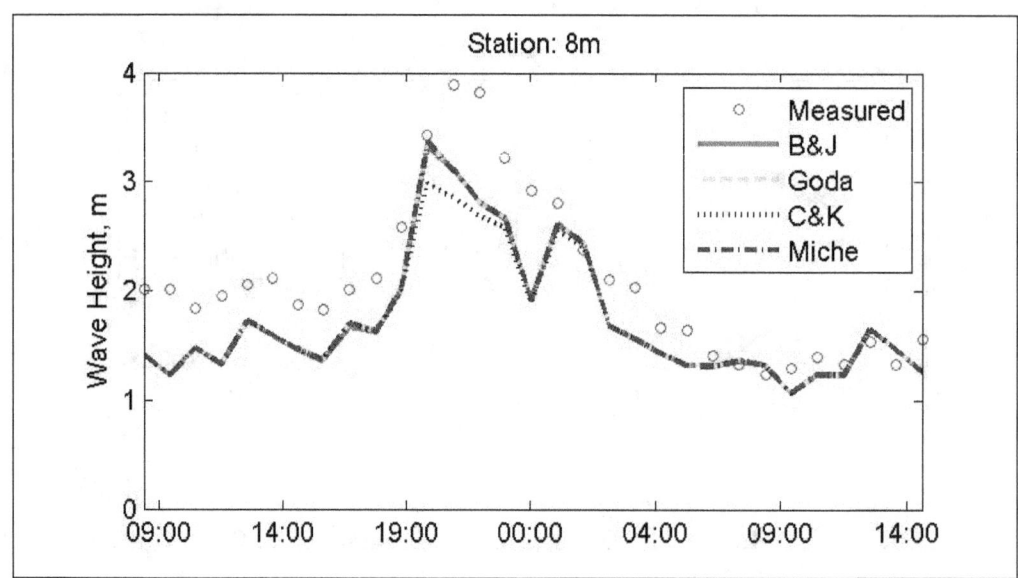

Figure 52. Calculated and measured wave height at 8-m station for Event 4, 22-23 August 2008.

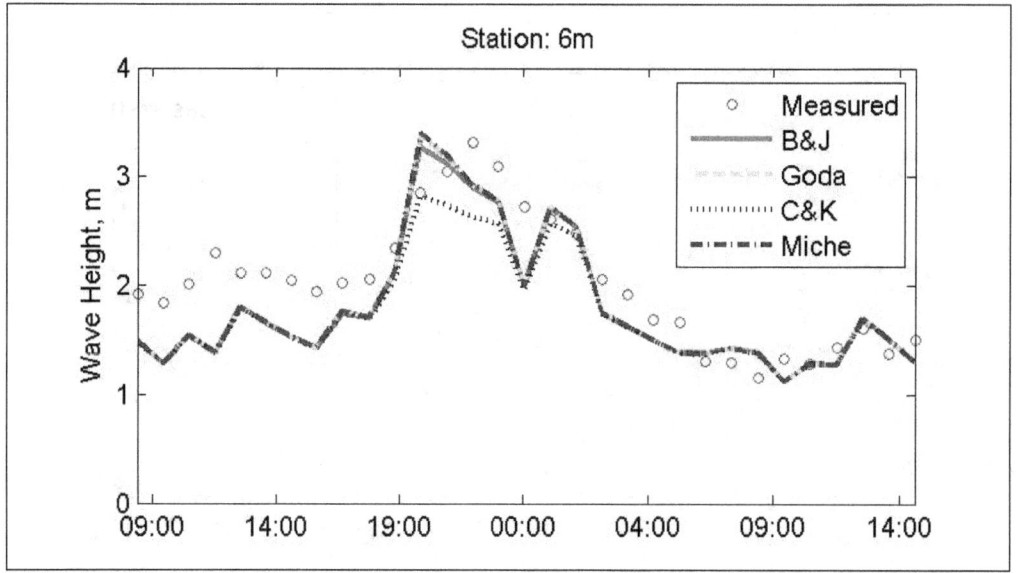

Figure 53. Calculated and measured wave height at 6-m station for Event 4, 22-23 August 2008.

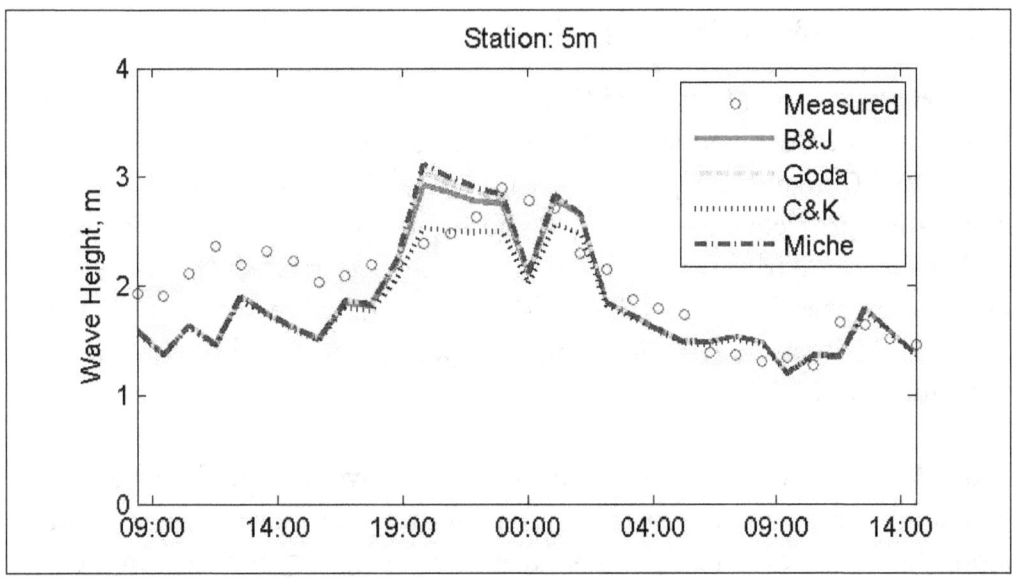

Figure 54. Calculated and measured wave height at 5-m station for Event 4, 22-23 August 2008.

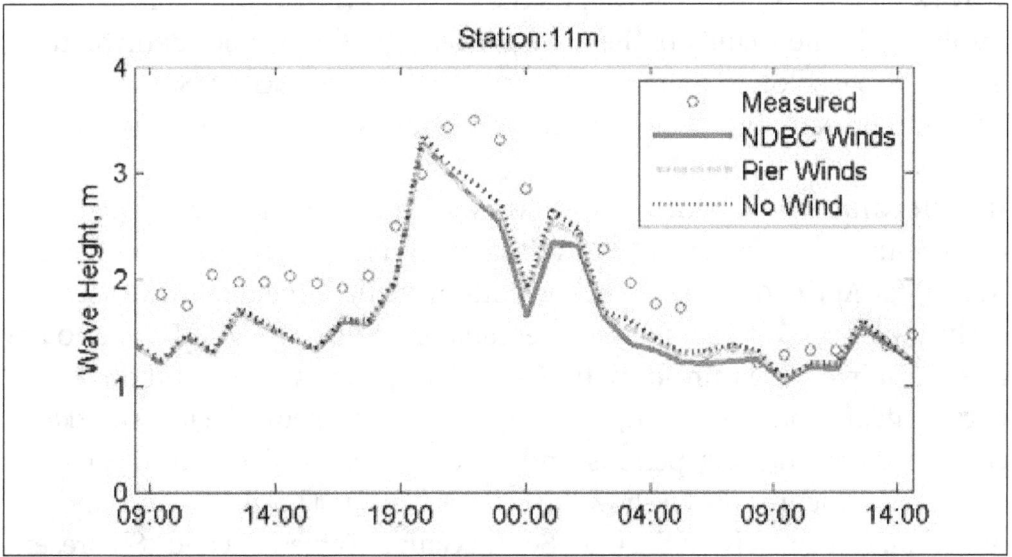

Figure 55. Measured and calculated wave height at 11-m station with various wind forcing.

CMS-Wave with the default values of parameters varied for this case provided the best result as compared to data in the storm wave simulations at the FRF. For a relatively small model domain, CMS-Wave was not sensitive to including the wind input in the simulation. Four different wave breaking formulations available in CMS-Wave all produced similar results. CMS-Wave results agreed better without including the bottom friction at the FRF. It is important to note that CMW-Wave is a steady state model; when waves are changing during an evolving storm like a hurricane, larger errors will be introduced into the model's results.

4.2.6   Test C3-Ex6: Mississippi Coastal Improvement Program

Description: The Mississippi Coastal Improvements Program (MsCIP) has maintained two nearshore directional wave gauges (COE Gulf Gauge and Sound Gauge) at Ship Island, MS, as part of the barrier island restoration project (USACE, 2010). These gauges measure wave height and period regularly. The wave direction is reported only if the wave height is greater than 0.1 m. The offshore wave data are available from a NDBC directional Buoy 42040 (165-m depth), located 90 km offshore Dauphin Island, AL.

Model Setup and Parameters: The CMS-Wave rectangular grid extended 41 km from the 15-m depth contour to the shoreline and 93 km alongshore. The domain includes variable cell size of 5 to 200 m in the cross-shore direction and of 70 to 200 m in the alongshore direction. None of the grid cells exceeded a 5:1 aspect ratio, as recommended for CMS-Wave simulations. Figure 56 shows the CMS-Wave grid with higher resolution in the vicinity of the wave gauges and Ship Island. Bathymetric data were based on a combination of surveys by the USGS, USACE and NOAA, representing the most recent conditions (Wamsley et al. 2011).

Results and Discussion: CMS-Wave was run in standalone full-plane mode with input wind from NOAA Coastal Station 8744707 (Gulfport Outer Range) for April 2010 neglecting the bottom friction. Incident wave spectra at the model seaward boundary were transformed from NDBC Buoy 42040 using a simple wave transformation model, i.e., Snell's Law and shore parallel depth contour assumption. Figures 57 to 59 show the measured and calculated wave heights, periods, and directions, respectively, at the COE Gulf Gauge. Figures 60-62 show the comparisons for height, period, and direction, respectively, at the COE Sound Gauge. Tables 18a and 18b present the statistics for the measured and calculated wave heights, periods, and directions at Gulf Gauge and Sound Gauge, respectively. The calculated wave heights generally agree with data within 0.2m (30-40% RMSE), although large errors in gauge measurements are expected for such low wave heights. The statistics for the calculated wave periods and directions are generally not as good as wave height because the calculated wave period and direction parameters cannot represent the multiple-peaked or bi-modal wave conditions (such as local wind waves superposed over the swell). Figure 63 shows an example of multiple-peaked spectrum at NDBC 42040, 1500 GMT, 7 April 2010.

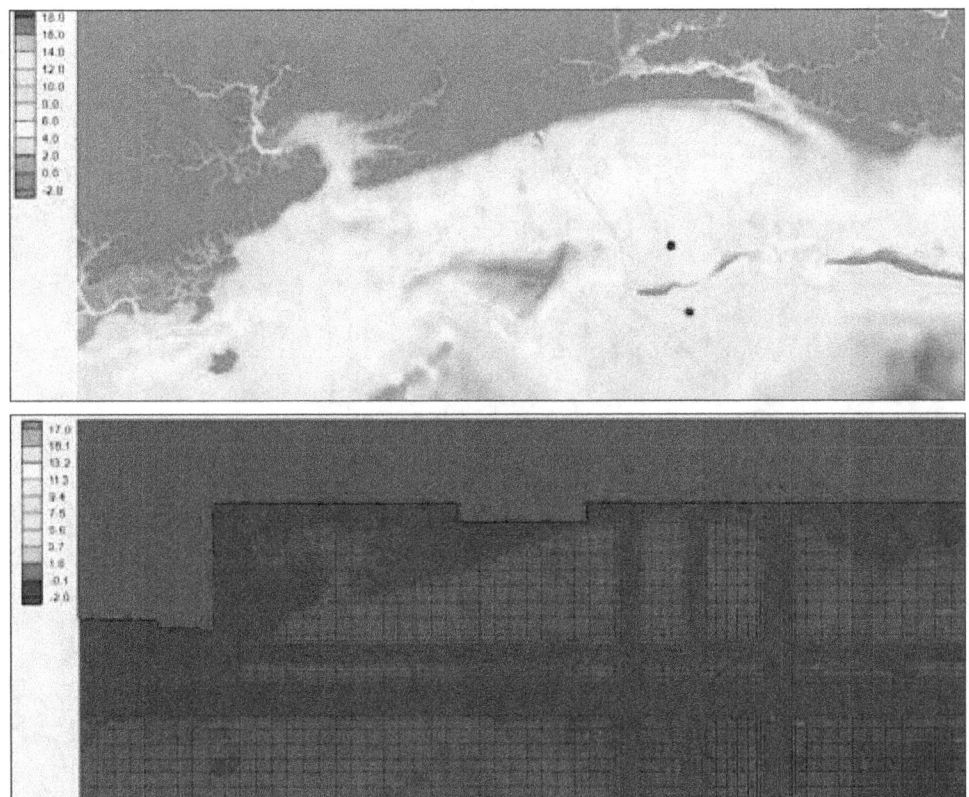

Figure 56. CMS-Wave grid domain (bottom) and two local wave gauge locations (top; black dots).

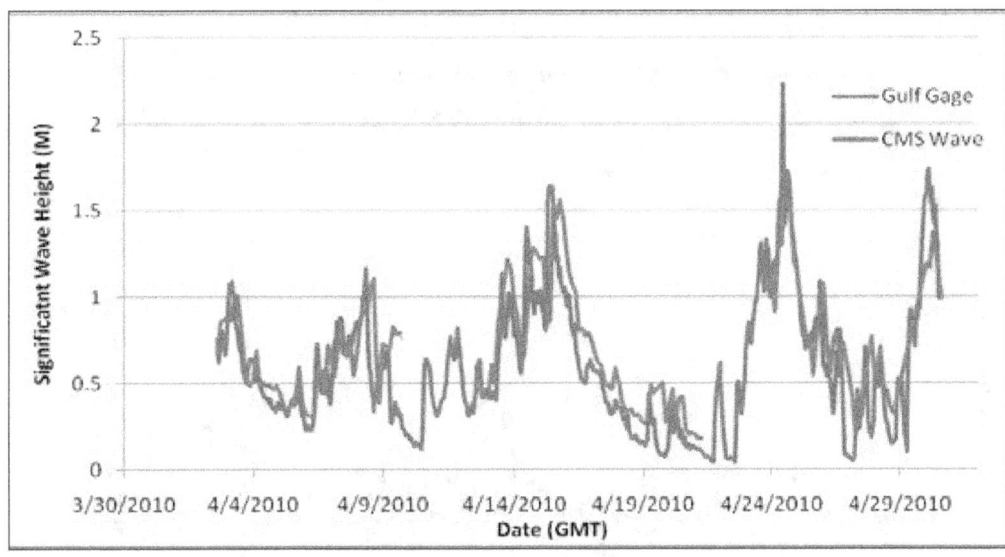

Figure 57. Measured and Calculated wave heights at COE Gulf gauge, April 2010.

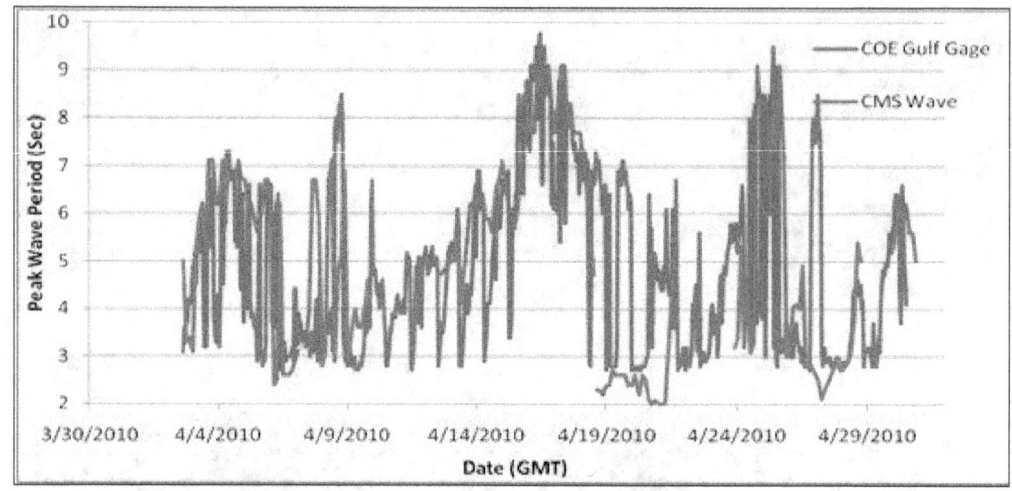

Figure 58. Measured and Calculated wave periods at COE Gulf gauge, April 2010.

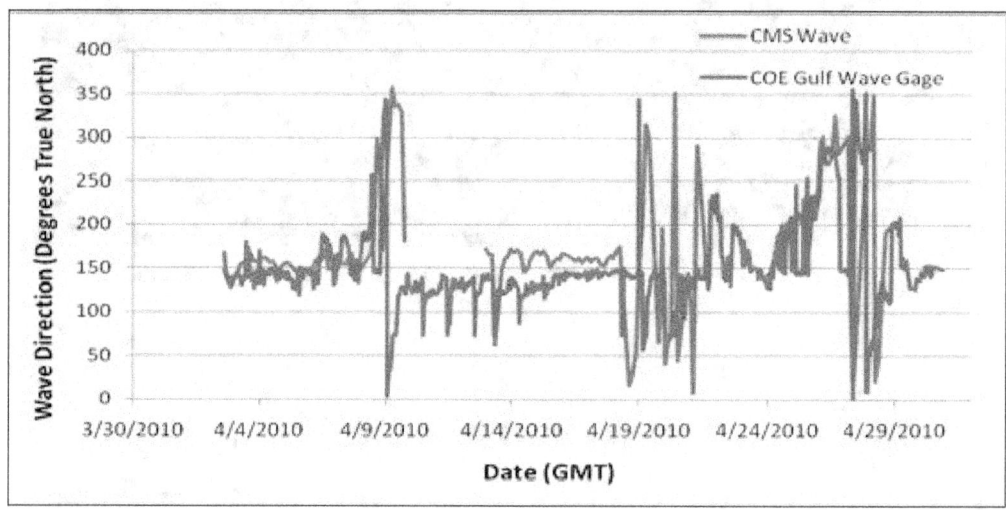

Figure 59. Measured and Calculated wave directions at COE Gulf gauge, April 2010.

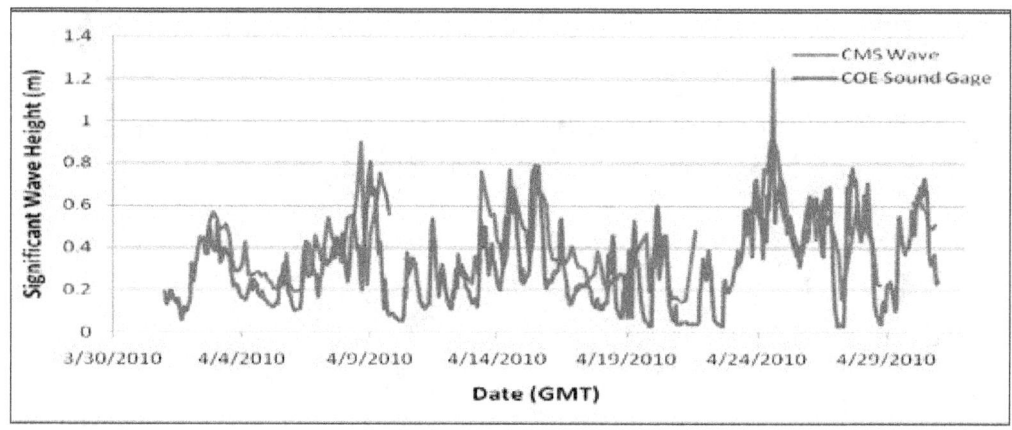

Figure 60. Measured and Calculated wave heights at COE Sound gauge, April 2010.

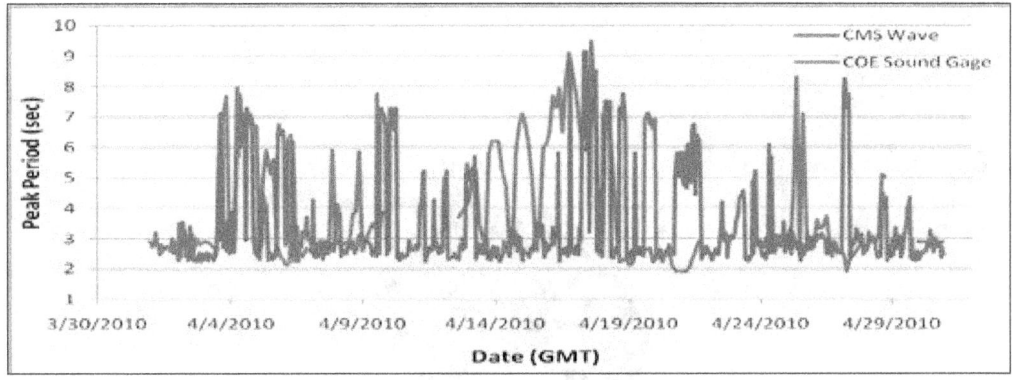

Figure 61. Measured and Calculated wave periods at COE Sound gauge.

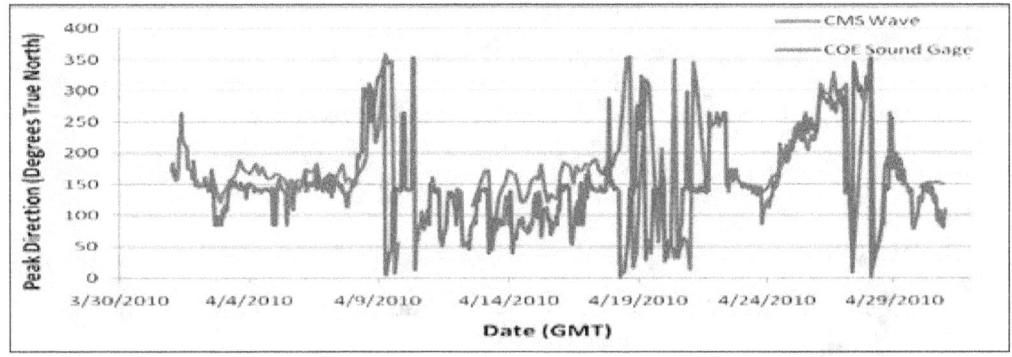

Figure 62. Measured and Calculated wave directions at COE Sound gauge, April 2010.

Table 18a. Wave Statistics at COE Gulf Gauge.

Statistics	H_s (m)	T_p (sec)	Direction (deg)
RMSE	0.231	2.03	78.3
R	0.85	0.54	0.28
MAE	0.42	1.21	6.8

Table 18b. Wave Statistics at COE Sound Gauge.

Statistics	H_s (m)	T_p (sec)	Direction (deg)
RMSE	0.173	2.43	92.8
R	0.71	0.18	0.35
MAE	0.36	1.30	7.6

The calculated wave height agreed better with data than the wave period and direction. Overall better agreement was obtained for the Gulf Gauge than the Sound Gauge with the island sheltering degrading the Sound Gauge comparisons. The model results were closer to data without the bottom friction at Ship Island comparing to field data. For relatively low wave heights and short propagation distances on a sandy bed environment, the bottom friction should not an important factor.

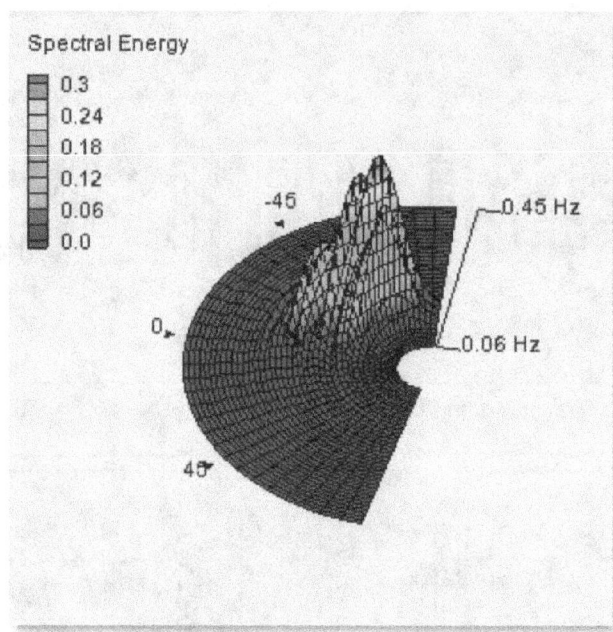

Figure 63. Multi-peak spectrum at NDBC 42040, 1500
GMT, 7 April 2010.

4.2.7   Test C3-Ex7: Waves over a submerged rock reef, Indian River County, FL

Description: Directional wave information was collected by Surfbreak Engineering Sciences, Inc. (SES, 2011) in Indian River County, Florida to quantify nearshore wave transformation over submerged rock reefs. An ADCP was installed offshore of the reef at the 9-m depth and an ADV was deployed inshore of the reef at 2-m to 3-m depth to measure current and waves. Figure 64 shows the location of the instruments and an aerial of the project site. A more detailed description of this field experiment is available, including specifics of instrumentation used and data analyses (SES, 2011).

Model Setup and Parameters: The CMS-Wave grid extended 980 m from the ADCP to the shoreline and 1,230 m alongshore, with a constant cell size of 25 m x 25 m. Figure 65 shows the CMS-Wave domain with the bathymetry based on data collected in 1997 and 2002 (SES 2011). The directional spectra measured at the ADCP location served as incident wave conditions. Figure 66 shows incident wave heights and periods at the seaward boundary. Wind measurements were available at Spessard Holland Park, located 45 km to the north. Water level was obtained from the ADCP. Wetting and drying, and diffraction with intensity of 4 were included in the simulation.

Figure 64. Vero Beach showing the location of deployed instrument.

Sensitivity to bottom friction was investigated by varying the Manning friction coefficient (n). The wave breaking criterion was based on the Battjes and Janssen formula. Three different bottom friction coefficients were tested: 0.0, 0.03 and 0.3. Bottom friction values are constant in space over the domain. Figure 67 shows measured significant wave height at the nearshore ADV compared to CMS-Wave results with various bottom friction coefficients. Figures 68 and 69 compare measured period and direction to the CMS-Wave results for n=0.3.

Results and Discussion: Tables 19a, 19b, and 19c present statistics for the measured and calculated wave heights, periods, and directions, respectively. The CMS-Wave simulation with Manning's n = 0.3, approximately ten times the typical value applied over the sandy bed, provides the best fit for wave height. However, wave period and direction calculations show no significant dependence on bottom friction.

CMS-Wave can approximate the wave transformation over the shallow reef if a large bottom friction is used in the simulations. Applying a large Manning bottom friction coefficient in CMS-Wave can produce sufficient damping to calculate the wave transformation over a reef bottom. More data for model calibration are required to assess similar mechanisms expected to induce damping. The Manning bottom friction coefficient must be used in CMS-Wave with high bottom friction coefficients.

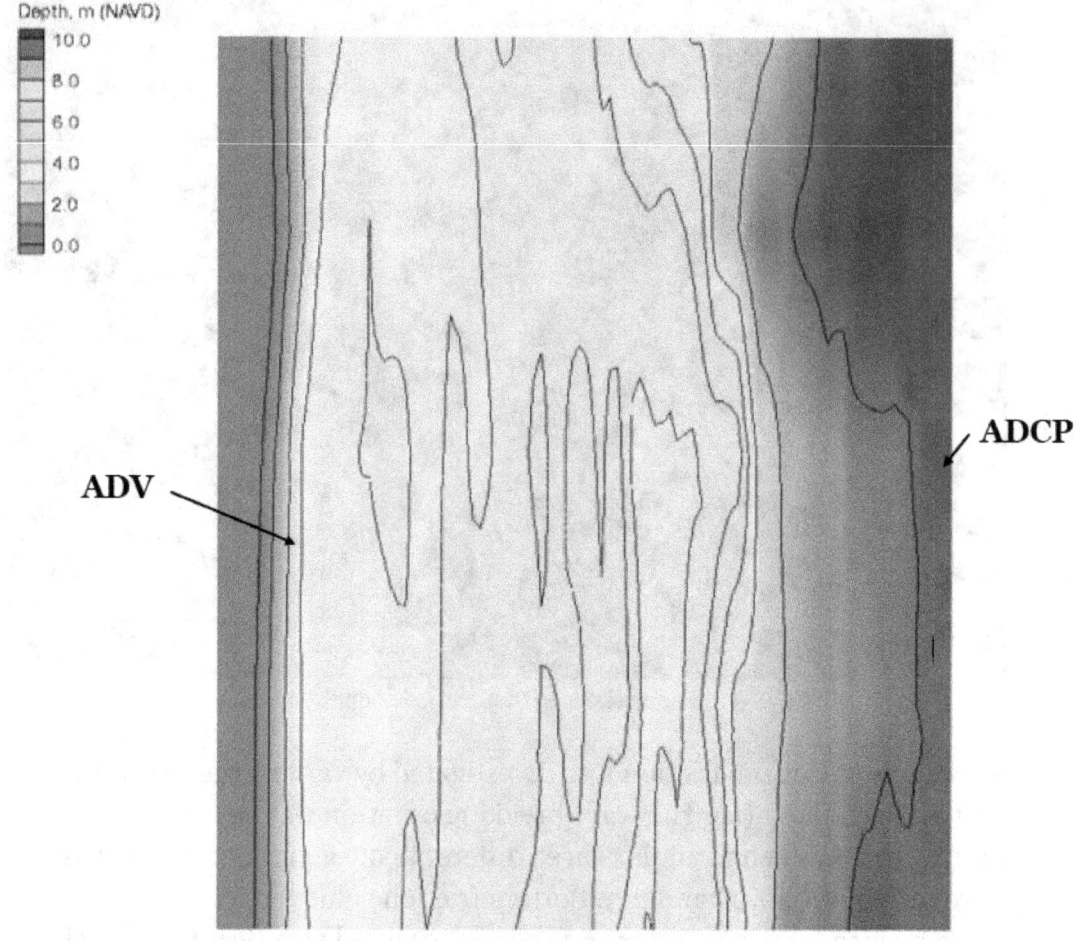

Figure 65. CMS-Wave grid bathymetry.

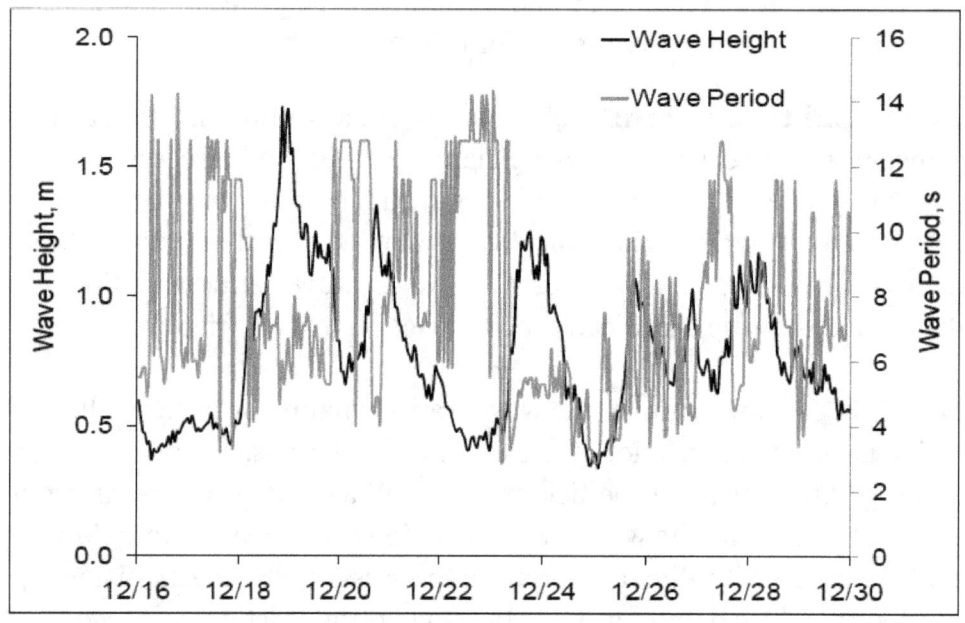

Figure 66. Wave forcing applied on the offshore boundary.

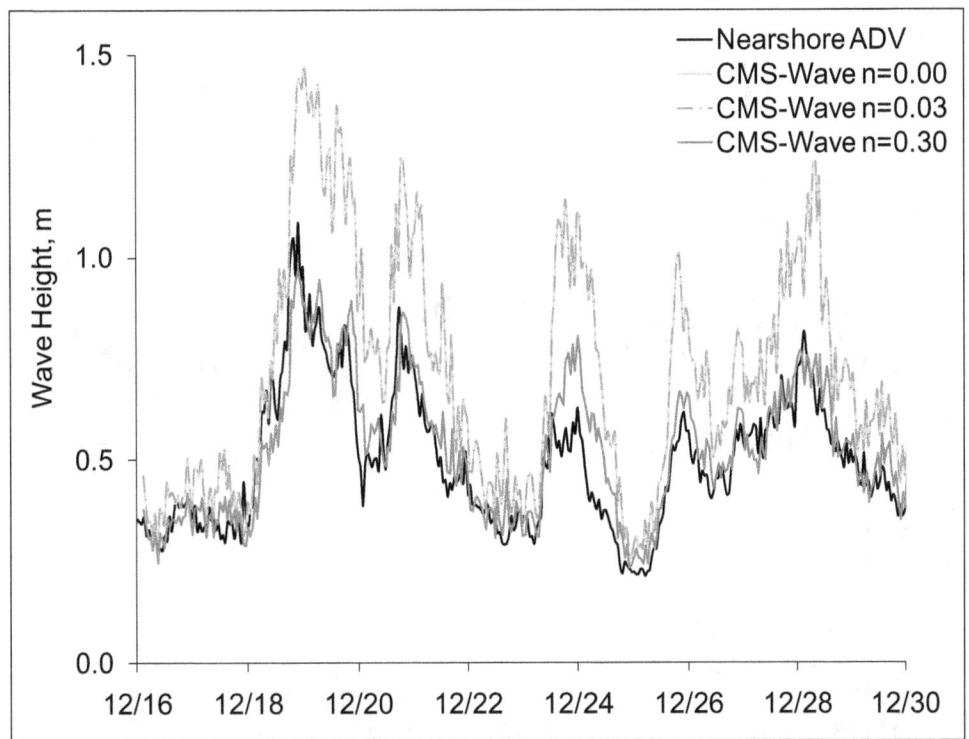

Figure 67. Measured and calculated wave heights at the nearshore ADV.

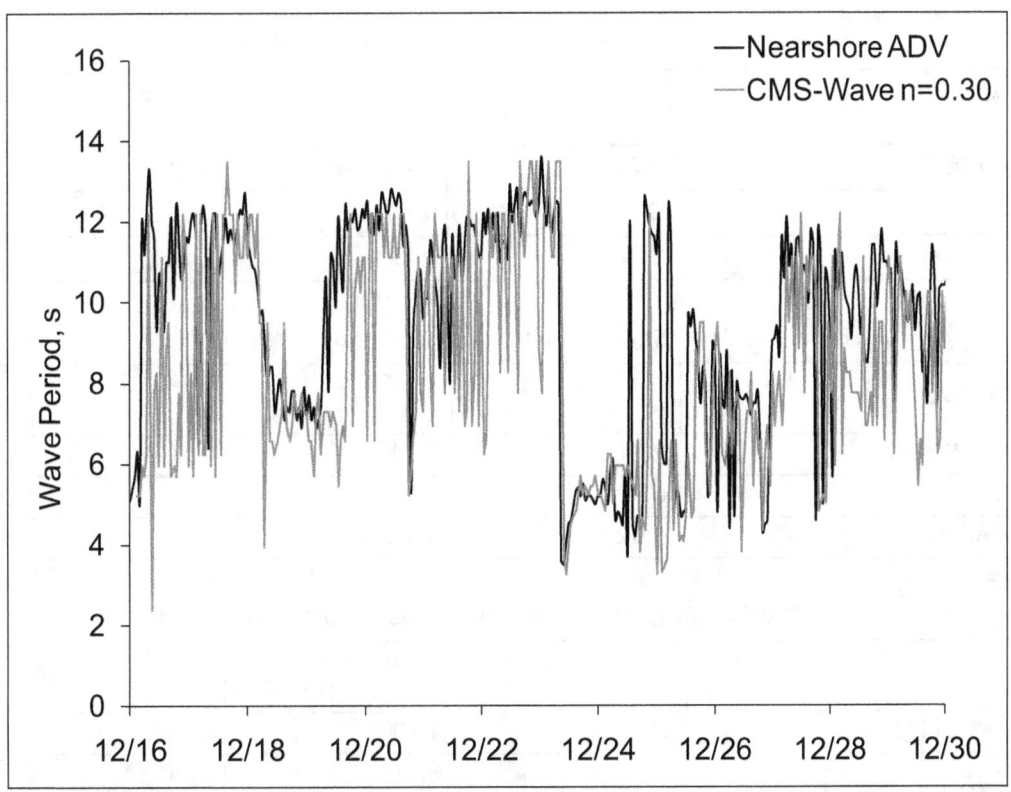

Figure 68. Measured and calculated wave periods at the nearshore ADV.

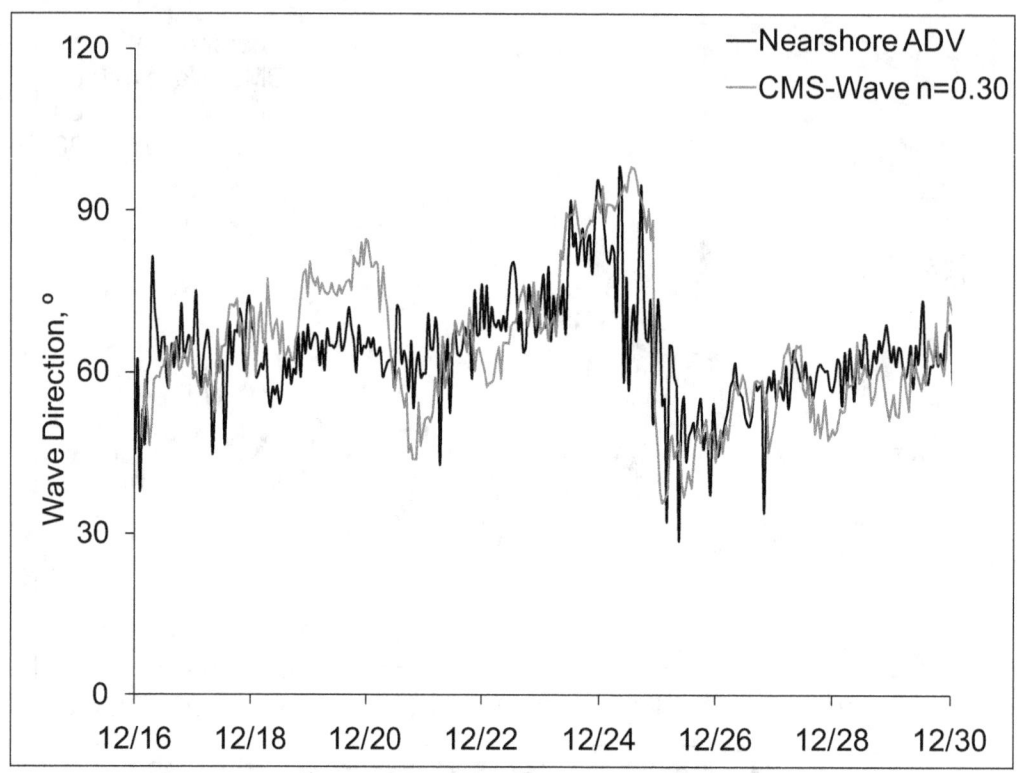

Figure 69. Measured and calculated wave directions at the nearshore ADV.

Table 19a. Sensitivity of wave height statistics to bottom friction.

Statistics	$n = 0.00$	$n = 0.03$	$n = 0.30$
RMSE (m)	0.27	0.27	0.08
R	0.77	0.77	0.81
MAE (m)	0.21	0.22	0.06

Table 19b. Sensitivity of wave period statistics to bottom friction.

Statistics	$n = 0.00$	$n = 0.03$	$n = 0.30$
RMSE (sec)	2.74	2.74	2.81
R	0.32	0.32	0.30
MAE (sec)	1.92	1.92	2.00

Table 19c. Sensitivity of wave direction statistics to bottom friction.

Statistics	$n = 0.00$	$n = 0.03$	$n = 0.30$
RMSE (deg)	10.3	10.3	10.5
R	0.41	0.41	0.40
MAE (deg)	7.91	7.90	8.14

5 Summary and Recommendations

Results of a comprehensive V&V study were described in this report to assess features and capabilities of the CMS-Wave model. Test cases included analytical/empirical solutions, idealized applications, and a number of studies with data from laboratory and field measurements. Applications considered in this V&V study included coastal inlets, navigation channels, coastal structures, bays, estuaries, barrier islands, and fringing reefs. The CMS-Wave model setup was described for each of these tests, including grid specifications and input parameters, and results were discussed with recommendations for practical applications. Both default values of key computational parameters and non-standard input parameters used in these simulations were described, and recommendations were provided for similar practical applications. Limitations of the model have been identified in the discussion of results for each test case.

A number of additional test applications are under investigation and these will be provided to users in future companion reports. Those test cases which have been completed and included in this report provide users with useful guidelines for applying the CMS-Wave model to similar problems. The model performance evaluation metrics provided for the test cases can be used by other numerical models users as benchmark standards.

The input and output files of the numerical model and data used in the completed test cases have been posted to the CIRP website (http://cirp.usace.army.mil/) to benefit the coastal community at large. This unique resource will be available to ERDC researchers and academicians for future evaluation of numerical wave models. Feedback on any aspect of this V&V study from the user community is welcome.

5.1   Major findings

5.1.1   Category 1: Analytical solutions and idealized problems

- CMS-Wave was tested against SPM/CEM and proved to be reliable for wave generation and growth for coastal applications. This capability of model needs comparisons to field measurements.
- The capability of wave-wave interaction calculation was shown to be robust and it improves model prediction in the large scale coastal inlet applications.
- Wave diffraction calculations at a breakwater gap approximated SPM/CEM monographs for engineering applications, but phase-resolving models should be used in projects where diffraction is a key process.

5.1.2   Category 2: Laboratory studies with data

- CMS-Wave was validated for wave and ebb tidal current interactions in an idealized inlet physical model. The Battjes and Janssen wave breaking formula produced the overall best results, so it is recommended as the default for all cases and not just currents.
- For the test cases investigated, wave-induced longshore current on a planar beach showed weak a effect on wave breaking in the surf zone.
- Wave runup calculations for several sloping structures and different wave conditions agreed with laboratory measurements. Runup calculations performed better for flat slopes (less than 1:5).
- The model was validated for combined wave-current-structure interactions at Cleveland Harbor. The model result was sensitive to the strong discharge (river flow) from the Cuyahoga River.

5.1.3   Category 3: Field studies with data

- Full and half plane capabilities of the model were validated with data from Matagorda Bay, Texas, which is a large shallow bay (~ 2 m depth). The bi-modal wave system that existed in the measurements could not be represented by calculation of single wave height and period parameters. Full plane was necessary to simulate combined local wave generation in the bay and waves entering the bay from the Gulf of Mexico. The full-plane model required more computational resources for this application.
- Combined wave shoaling, reflection, refraction, and diffraction were validated with data for a navigation improvement project at Grays

Harbor and Half Moon Bay, Washington. Water level variation had the most effect on calculated waves and currents nearshore. Model comparisons in inner Half Moon Bay, where diffraction is critical, showed comparatively higher error in wave height, period and direction.

- Wave-current interactions were validated for the high-energy environment including a navigation channel and jetties at the Mouth of the Columbia River, Oregon/Washington. Wave sheltering and diffraction effects are strong at the North Jetty protecting large waves from the northwest. Again, model results were less accurate in the sheltered diffraction zone than in the more exposed parts of the inlet.

- The wave transformation feature was validated for a rough reef protecting Southeast Oahu coast, Hawaii. Large bottom friction coefficients were essential for accurate wave prediction. Calibration of the model with field data was required to get accurate results.

- CMS-Wave simulation of Atlantic storms and Hurricane Bill showed wave height variation in the cross-shore array at the FRF, North Carolina. Wave heights tended to be over predicted for the northeasters and underpredicted for the Hurricane Bill swell.

- The full plane capability was validated with field measurements for barrier island breaching and restoration at the Mississippi Coast. Calculation of single wave height and period parameters could not represent the bi-modal wave system that existed in the measurements.

- CMS-Wave was validated with field data along a rocky coast at Indian River County, Florida. The model behavior was sensitive to bottom friction coefficients and requires calibration.

- The absolute modeling error depends on the applications and complexity of processes involved. In general, the modeling error shall be considered in relative terms based on the physics included or alternatives considered. Wave modeling that does not involve wave-phase information is reliable for wave transformation in the open coast or wave generation in a large bay. In the vicinity of structures and in the surf zone where wave phase plays an important role, data are needed to calibrate the model or a phase-resolving model can be more adequate to use.

References

Ahrens, J. P., and M. S. Heimbaugh. 1988. Approximate upper limit of irregular wave runup on riprap. *Coastal Engineering Research Center Technical Report CERC-88-5*. Vicksburg, MS: U.S. Army Engineer Waterways Experiment Station.

Ahrens, J. P., and M. F. Titus. 1981. Laboratory data report: irregular wave runup on plane smooth slopes. *Coastal Engineering Research Center unpublished Laboratory Report*. Vicksburg, MS: U.S. Army Engineer Waterways Experiment Station.

Battjes, J. A. 1972. Set-up due to irregular waves. *Proceedings 13th International Conference on Coastal Engineering*, ASCE, 1993-2004.

Battjes, J. A., and J. Janssen. 1978. Energy loss and set-up due to breaking of random waves. *Proceedings 16th International Conference Coastal Engineering*, ASCE, 569-587.

Booij, N., R. C. Ris, and L. H. Holthuijsen. 1999. A third-generation wave model for coastal regions: 1. Model description and validation. *Journal of Geophysical Research* 104(C4):7,649-7,666.

Bottin, Jr. R.R. 1983. Cleveland Harbor, Ohio, design for the safe and efficient passage of 1000-ft-long vessels at the west (main) entrance. *Hydraulics Laboratory Technical Report HL-83-6*. Vicksburg, MS: U.S. Army Engineer Waterways Experiment Station.

Bouws, E., H. Gunther, W. Rosenthal, and C.L. Vincent. 1985. Similarity of the wind wave spectrum in finite depth water: 1. Spectral form. *Journal of Geophysical Research*, 90 C1: 975-86.

Buttolph, A. M., C. W. Reed, N. C. Kraus, N. Ono, M. Larson, B. Camenen, H. Hanson, T. Wamsley, and A. K. Zundel. 2006. Two-dimensional depth-averaged circulation model CMS-M2D: Version 3.0, Report 2: Sediment transport and morphology change. *Coastal and Hydraulics Laboratory Technical Report ERDC/CHL TR-06-09*. Vicksburg, MS: U.S. Army Engineer Research and Development Center.

Chawla, A., and J. T. Kirby. 2002. Monochromatic and random wave breaking at blocking points. *Journal of Geophysical Research 107(C7)*, 10.1029/2001JC001042.

Cialone, M. A., M. E. Brown, J. M. Smith, and K. K. Hathaway. 2008. Southeast Oahu Coastal Hydrodynamic Modeling with ADCIRC and STWAVE. *Coastal and Hydraulics Laboratory Technical Report ERDC/CHL TR-08-9*. Vicksburg, MS: U.S. Army Engineer Research and Development Center.

Coastal Engineering Manual, 2006. U.S. Army Corps of Engineers Headquarters, Washington, D.C.

Demirbilek, Z., L. Lin, and A. Zundel. 2007a. WABED model in the SMS: Part 2. Graphical interface. *Coastal and Hydraulics Laboratory Engineering Technical Note ERDC/CHL CHETN-I-74*. Vicksburg, MS: U.S. Army Engineer Research and Development Center.

Demirbilek, Z., O. Nwogu, and D. L. Ward. 2007b. Laboratory study of wind effect on runup over a fringing reef. *Coastal and Hydraulics Laboratory Technical Report ERDC/CHL TR-07-4*. Vicksburg, MS: U.S. Army Engineer Research and Development Center.

Demirbilek, Z., L. Lin, and O.G. Nwogu. 2008. Wave modeling for jetty rehabilitation at the Mouth of the Columbia River, Washington/Oregon, USA. *Coastal and Hydraulics Laboratory Technical Report ERDC/CHL TR-08-3*. Vicksburg, MS: U.S. Army Engineer Research and Development Center.

Demirbilek, Z., M. Mohr, and S. Chader, 2010. Phase 1 study final letter report: Wave modeling for Cleveland Harbor, Ohio, Letter Report, Jul 2010, p.77.

Demirbilek, Z., and V. Panchang. 1998. CGWAVE: A coastal surface-water wave model of the mild-slope equation. *Coastal and Hydraulics Laboratory Technical Report CHL-98-26*. Vicksburg, MS: U.S. Army Engineer Waterways Experiment Station.

Goda, Y. 1970. A synthesis of breaker indices. *Transactions of the Japan Society of Civil Engineers* 13:227-230 (in Japanese).

Goda, Y. 1985. Random seas and design of maritime structures. Tokyo: University of Tokyo Press.

Goda, Y. 2006. Examination of the influence of several factors on longshore current computation with random waves. *Coastal Engineering* 53(2-3):157-170.

Hanson, J.L., Friebel, H.C., and Hathaway, K.K. 2009. Coastal Wave Energy Dissipation: Observations and STWAVE-FP performance. 11TH International Workshop on Wave Hindcasting and Forecasting & 2nd Coastal Hazards Symposium; Halifax, Nova Scotia, Canada; October 18-23, 2009.

Hasselmann, K., T. P. Barnett, E. Bouws, H. Carlson, D.E. Cartwright, K. Enke, J.A. Ewing, H. Gienapp, D.E. Hasselmann, P. Kruseman, A. Meerbrug, P. Muller, D. J. Olbers, K. Richter, W. Sell, and H. Walden. 1973. Measurements of wind-wave growth and swell decay during the Joint North Sea Wave Project (JONSWAP). *Deutsche Hydrographische Zeitschrift A80(12)*, 95 p.

Hasselmann, S., K. Hasselmann, J.H. Allender, and T.P. Barnett. 1985. Computations and parameterizations of the nonlinear energy transfer in a gravity wave spectrum. Part II. Parameterizations of the nonlinear energy transfer for application in wave models. *Journal of Physical Oceanography* 15:1378-1391.

Hamilton, D.G., and B. A. Ebersole. 2001. Establishing uniform longshore currents in a large-scale laboratory facility. *Coastal Engineering*, 42, 199-218

Headquarters, U.S. Army Corps of Engineers. 2002. *Coastal Engineering Manual. EM 1110-2-1100*. Washington, DC (in 6 volumes).

HMB, 2011. http://cirp.usace.army.mil/news/CIRP_News/CIRP_eNewsletter_Jun2011.pdf.

HSC, 2010. http://cirp.usace.army.mil/news/CIRP_News/CIRP-news-Mar10.html.

Jenkins, A. D., and O. M. Phillips. 2001. A simple formula for nonlinear wave-wave interaction. *International Journal of Offshore and Polar Engineering* 11(2):81-86.

Johnson, J. W. 1952. Generalized wave diffraction diagrams. *Proceedings 2nd Conference on Coastal Engineering*, ASCE.

Lin, L., and Z. Demirbilek. 2005. Evaluation of two numerical wave models with inlet physical model. *Journal of Waterway, Port, Coastal, and Ocean Engineering* 131(4):149-161, ASCE.

Lin, L., Z. Demirbilek, and F. Yamada. 2008. CMS-Wave: A nearshore spectral wave processes model for coastal inlets and navigation projects. *Coastal and Hydraulics Laboratory Technical Report ERDC/CHL TR-08-13*. Vicksburg, MS: U.S. Army Engineer Research and Development Center.

Lin, L., Z. Demirbilek, J. Zheng, and H. Mase. 2010. Rapid calculation of nonlinear wave-wave interactions in wave-action balance equation. *Proceedings of the International Conference on Coastal Engineering, No. 32*. Shanghai, China. Retrieved from http://journals.tdl.org/ICCE/.

Lin, L., Z. Demirbilek, and H. Mase. 2011. Recent capabilities of CMS-Wave: A coastal wave model for inlets and navigation projects. *Proceedings, Symposium to honor Dr. Nicholas Kraus. Journal of Coastal Research, Special Issue 59,7-14*.

Longuet-Higgins, M. S., and R. W. Stewart. 1961. The changes in amplitude of short gravity waves on steady non-uniform currents. *Journal of Fluid Mechanics* 10(4):529-549.

Mase, H., and Y. Iwagaki. 1984. Runup of random waves on gentle slopes. *Proceedings 19th International Conference on Coastal Engineering*, ASCE, 593-609.

Mase, H. 1989. Random wave runup height on gentle slope. *Journal of Waterway, Port, Coastal, and Ocean Engineering* 85(3):123-152, ASCE.

Mase, H. 2001. Multidirectional random wave transformation model based on energy balance equation. *Coastal Engineering Journal* 43(4):317-337 JSCE.

Mase, H., H. Amamori, and T. Takayama. 2005a. Wave prediction model in wave-current coexisting field. *Proceedings 12th Canadian Coastal Conference* (CD-ROM).

Mase, H., K. Oki, T. S. Hedges, and H. J. Li. 2005b. Extended energy-balance-equation wave model for multidirectional random wave transformation. *Ocean Engineering* 32(8-9):961-985.

Miche, M. 1951. Le pouvoir reflechissant des ouvrages maritimes exposes a 1'action de la houle. *Annals des Ponts et Chau.ssess. 121e Annee*: 285-319 (translated by Lincoln and Chevron, University of California, Berkeley, Wave Research Laboratory, Series 3, Issue 363, June 1954).

Militello, A., C. W. Reed, A. K. Zundel, and N. C. Kraus. 2004. Two-dimensional depth-averaged circulation model CMS-M2D: Version 2.0, Report 1, Technical documentation and user's guide. *Coastal and Hydraulics Laboratory Technical Report ERDC/CHL TR-04-02*. Vicksburg, MS: U.S. Army Engineer Research and Development Center.

Moritz, H. R. 2005. Mouth of the Columbia River mega-transect instrument deployment, internal document. Portland, Oregon: U.S. Army Engineer District, Portland.

NH, 2009. http://cirp.usace.army.mil/news/CIRP_News/CIRP-news-Dec09.html.

NOAA, 2010. http://tidesandcurrents.noaa.gov/ports/index.shtml?port=gp

Nwogu, O., and Z. Demirbilek. 2001. BOUSS-2D: A Boussinesq wave model for coastal regions and harbors. *Coastal and Hydraulics Laboratory Technical Report ERDC/CHL TR-01-25*. Vicksburg, MS: U.S. Army Engineer Research and Development Center.

Osborne, P. D., and M. H. Davies. 2004. South jetty sediment processes study, Grays Harbor, Washington: Processes along Half Moon Bay. PIE Technical Report.Sakai, S., N. Kobayashi, and K. Koike. 1989. Wave breaking criterion with opposing current on sloping bottom: an extension of Goda's breaker index. *Annual Journal of Coastal Engineering* 36:56-59, JSCE (in Japanese).

Puckette, T. 2006. Matagorda Bay Field Data Collection. Evans-Hamilton Inc. Draft Report. EHI Project No. 5503.

Seabergh, W. C., W. R. Curtis, L. J. Thomas, and K. K. Hathaway. 2002. Physical model study of wave diffraction-refraction at an idealized inlet. *Coastal Inlet Research Program Technical Report ERDC/CHL-TR-02-27*. Vicksburg, MS: U.S. Army Engineer Research and Development Center.

Seabergh, W. C., L. Lin, and Z. Demirbilek. 2005. Laboratory study of hydrodynamics near absorbing and fully reflecting jetties. *Coastal Inlets Research Program, Technical Report ERDC/CHL-TR-05-8*. Vicksburg, MS: U.S. Army Engineer Research and Development Center.

Shore Protection Manual, 1984. 4th ed., 2 Vol, U.S. Army Engineer Waterways Experiment Station, U.S. Government Printing Office, Washington, DC.

Smith, J. M. 2001. Modeling nearshore transformation with STWAVE. *Coastal and Hydraulics Laboratory Special Report ERDC/CHL SR-01-01*. Vicksburg, MS: U.S. Army Engineer Research and Development Center.

Smith, J. M., W. C. Seabergh, G. S. Harkins, and M. J. Briggs. 1998. Wave breaking on a current at an idealized inlet. *Coastal and Hydraulics Laboratory Technical Report CHL-98-31*. Vicksburg, MS: U.S. Army Engineer Waterways Experiment Station.

Surfbreak Engineering Sciences (SES). 2011. Field Data for Testing CMS-Wave in Regard to Frictional Energy Losses Induced by Low-Relief Rock Reef: R-68 IRC. Contract Report for the U.S. Army Coastal and Hydraulics Laboratory, Engineer Research and Development Center. March 7,2011.

USACE. 2010. US Army Corps of Engineers, Mississippi Coastal Improvements Program. http://www.sam.usace.army.mil/mscip/default.html.

Visser, P. J. 1991. Laboratory measurements of uniform longshore currents. *Coastal Engineering* 15:563-593.

Wamsley, T., B. W. Bunch, R. S. Chapman, M. B. Gravens, A. S. Grzegorzewski, B. D. Johnson, R. L. Permenter, and M. W. Tubman. 2011. Mississippi Coastal Improvement Program, Barrier Island Restoration Numerical Modeling. In Progress. *Technical Report*, Vicksburg, MS: U.S. Army Engineer Research and Development Center.

Wiegel, R. L. 1962. Diffraction of waves by a semi-infinite breakwater. *Journal of the Hydraulics Division* 88(HY1):27-44.

Appendix A: Summary of CMS-Wave Control Parameters with Default Values

These latest versions have the optional full-plane capability for users. The *.std can have a max of 25 parameters - the first 15 parameters are defined the same as in the CMS-Wave TR, the other 10 parameters are new.

1st	2nd	3rd	4th	5th	6th
iprp	icur	ibk	irs	kout	ibnd

7th	8th	9th	10th	11th	12th	13th	14th	15th
iwet	ibf	iark	iarkr	akap	bf	ark	arkr	iwvbk

16th	17th	18th	19th	20th	21st	22nd	23rd	24th	25th
nonln	igrav	irunup	imud	iwnd	isolv	ixmdf	iproc	iview	iroll

At least the first 6 parameters are needed in the *.std and the remaining parameters starting any parameter after the 6th will be assigned to default values if not specified in *.std. The description of the 1st to 24th parameters is given below.

iprp = 0 (wave propagation with wind input in *.eng)
     1 (wave propagation only, neglect wind input in *.eng)
     -1 (fast mode)
     2 (forced grid internal rotation)
     3 (without lateral energy flux)

icur = 0 (no current input)
     1 (with current input *.cur)
     2 (with *.cur, use only the 1st set current data)

ibk = 0 (no wave break info output)
      1 (output breaking indices *.brk)
      2 (output energy dissipation rate *.brk)

irs = 0 (no wave radiation stress calc)
      1 (output radiation stress *.rad)
      2 (calculate/output setup/max-water-level + *.rad)

kout = number of special wave output location, output spectrum in *.obs
      and parameters in selhts.out

ibnd = 0 (no input a parent spectrum *.nst)
      1 (read *.nst, averaging input spectrum)
      2 (read *.nst, spatially variable spectrum input)

iwet = 0 (allow wet/dry, default)
      1 (without wet/dry)
      -1 (allow wet/dry, output swell and local sea files)
      -2 (output combined steering wav files)
      -3 (output swell, local sea, and combined wav files)

ibf = 0 (no bottom friction calc)
      1 (constant Darcy-Weisbach coef, c_f)
      2 (read variable c_f file, *.fric)
      3 (constant Mannings n)
      4 (read variable Mannings n file, *.fric)

iark = 0 (without forward reflection)
      1 (with forward reflection)

iarkr = 0 (without backward reflection)
      1 (with backward reflection)

akap = 0 to 4 (diffraction intensity, 0 for zero diffraction,
      4 for strong diffraction, default)

bf =   constant bottom friction coef c_f or n
      (typical value is 0.005 for c_f and 0.025 for Mannings n)

ark = 0 to 1 (constant forward reflection coef, global specification,
0 for zero reflection, 1 for 100% or fully reflection)

arkr = 0 to 1 (constant backward reflection coef, global specification,
0 for zero reflection, 1 for 100% or fully reflection)

iwvbk = 0 to 3 (option for the primary wave breaking formula:
0 for Goda-extended, 1 for Miche-extended,
2 for Battjes and Janssen, 3 for Chawla and Kirby)

nonln = 0 (none, default) 1 (nonlinear wave-wave interaction)

igrav = 0 (none, default) 1 (infra-gravity wave enter inlets)

irunup = 0 (none, default) 1 (automatic, runup relative to absolute datum)
2 (automatic, runup relative to updated MWL)

imud = 0 (mud.dat, default) 1 (none) -------- need it for users
who may not want to
include mud effect
as the mud.dat exists
(typical max kinematic
viscosity in mud.dat
is 0.04 m*m/sec)

iwnd = 0 (wind.dat, default) 1 (none) -------- need it in steering
if users decide not
to use the wind field
input when the wind
file exists

isolv = 0 (GSR solver, default) 1 (ADI)

ixmdf = 0 (output ascii, default) 1 (output xmdf) 2 (input & output xmdf)

iproc = 0 (same as 1, default) n (n processors for isolv = 0)
--- approx. processor
number=(total row)/300

iview = 0 (half-plane, default) 1 (full-plane) --- for the full plane, users can provide the additional input wave spectrum file wave.spc (same format as the *.eng) along the opposite side boundary (an imaginary origin for this wave.spc at the opposite corner; users can use SMS to rotate the CMS-Wave grid 180 deg to generate this wave.spc)

iroll = 0 to 4 (wave roller effect, 0 for no effect, default 4 for strong effect) -- more effective for finer resultion in the surf zone, say, for the cross-shore spacing < 10 m

REPORT DOCUMENTATION PAGE

Form Approved
OMB No. 0704-0188

1. REPORT DATE (DD-MM-YYYY) December 2011	2. REPORT TYPE Report 2 of a series	3. DATES COVERED (From - To)

4. TITLE AND SUBTITLE	5a. CONTRACT NUMBER
Verification and Validation of the Coastal Modeling System, Report 2: CMS-Wave	5b. GRANT NUMBER
	5c. PROGRAM ELEMENT NUMBER

6. AUTHOR(S)	5d. PROJECT NUMBER
Lihwa Lin, Zeki Demirbilek, Rob Thomas, and James Rosati, III	5e. TASK NUMBER
	5f. WORK UNIT NUMBER

7. PERFORMING ORGANIZATION NAME(S) AND ADDRESS(ES)	8. PERFORMING ORGANIZATION REPORT NUMBER
U.S. Army Engineer Research and Development Center Coastal and Hydraulics Laboratory 3909 Halls Ferry Road Vicksburg, MS 39180-6199	ERDC/CHL TR-11-10

9. SPONSORING / MONITORING AGENCY NAME(S) AND ADDRESS(ES)	10. SPONSOR/MONITOR'S ACRONYM(S)
	11. SPONSOR/MONITOR'S REPORT NUMBER(S)

12. DISTRIBUTION / AVAILABILITY STATEMENT

Approved for public release. Distribution is unlimited.

13. SUPPLEMENTARY NOTES

14. ABSTRACT

There are four reports documenting the Verification and Validation (V&V) of the Coastal Modeling System (CMS): an executive summary, waves, circulation, and sediment transport/morphodynamics, respectively. This is the second technical report (Report 2) that describes the wave modeling component of the V&V study. The goal of the report was to critically assess both general and special predictive skills of CMS-Wave, a spectral wave model in the CMS developed to address a variety of needs for coastal inlet applications. For model verification, a number of simple and idealized cases were selected to approve the basic physics and computational algorithms implemented in CMS-Wave. For model validation, a collection of more complicated cases with either laboratory or field data representing real world problems were assembled to confirm the overall performance or special capabilities of CMS-Wave. Provided in this report are descriptions of the V&V cases, model set up and boundary conditions specified in each case, and assessment of model performance. Major findings for each case are provided as guidance to users for future applications of CMS-Wave.

15. SUBJECT TERMS		
Circulation Coastal modeling system	Performance of models Sediment transport modeling	Verification and validation Waves

16. SECURITY CLASSIFICATION OF:			17. LIMITATION OF ABSTRACT	18. NUMBER OF PAGES	19a. NAME OF RESPONSIBLE PERSON
a. REPORT UNCLASSIFIED	b. ABSTRACT UNCLASSIFIED	c. THIS PAGE UNCLASSIFIED		111	19b. TELEPHONE NUMBER (include area code)

Standard Form 298 (Rev. 8-98)
Prescribed by ANSI Std. 239.18

www.ingramcontent.com/pod-product-compliance
Lightning Source LLC
Chambersburg PA
CBHW081239180526
45171CB00005B/472